Cosmological Special Relativity

COSMOLOGICAL SPECIAL RELATIVITY

The Large-Scale Structure of
Space, Time and Velocity

Second Edition

Moshe Carmeli

Ben Gurion University, Israel

Published by
World Scientific Publishing Co. Pte. Ltd.
P O Box 128, Farrer Road, Singapore 912805
USA office: Suite 1B, 1060 Main Street, River Edge, NJ 07661
UK office: 57 Shelton Street, Covent Garden, London WC2H 9HE

British Library Cataloguing-in-Publication Data
A catalogue record for this book is available from the British Library.

COSMOLOGICAL SPECIAL RELATIVITY
The Large-Scale Structure of Space, Time and Velocity (Second Edition)

Copyright © 2002 by World Scientific Publishing Co. Pte. Ltd.

All rights reserved. This book, or parts thereof, may not be reproduced in any form or by any means, electronic or mechanical, including photocopying, recording or any information storage and retrieval system now known or to be invented, without written permission from the Publisher.

For photocopying of material in this volume, please pay a copying fee through the Copyright Clearance Center, Inc., 222 Rosewood Drive, Danvers, MA 01923, USA. In this case permission to photocopy is not required from the publisher.

ISBN 981-02-4936-5

Printed in Singapore by Uto-Print

To *Eli, Dorith and Yair*

Preface

The study of cosmology in recent years became one of the most important and popular subjects. Cosmological theory connects different fields of research in physics, from elementary particles to the large-scale structure of the Universe. An example of this is the well-known inflationary universe model which emerged out of research in high-energy physics. The many experimental results obtained in the last years have reached the level of enabling us to determine what kind of Universe we live in.

The mathematical foundations of modern cosmology theory has a history that goes back to Einstein's relativity theory and is heavily based on it. It gives a successful description to the very important discovery of the expansion of the Universe by E. Hubble.

In this monograph we formulate cosmology, in the limit of *negligible* gravitational field, as a new special relativity describing the large structure of space, time and velocity in the Universe.

The theory unites space and velocity just as Einstein's special relativity theory does with space and time. Observers at different places in the Universe are subsequently related to each other by *relative* cosmic times just as those at different inertial systems are related by relative velocities in special relativity.

Subsequently, the group of cosmological transformations which relate distances and velocities at different cosmic times is derived, and the transformation is written down explicitly. This cosmological group of transformations has a similar structure to the familiar Lorentz group. Some important implications of the cosmological transformation are subsequently given.

A substantial part of the book is devoted to special relativity theory, which is presented along the lines of Einstein's original version. The stress is on the deep meaning of the theory rather than on technicalities, in order to emphasize the analogy to the cosmological special relativity. This part of the book, which is self-contained, gives a full exposition of Einstein's theory and can be used as a text for advanced undergraduate as well as graduate students.

The whole book is written in a clear and pedagogical way so as to enable nonexpert as well as expert readers to easily understand its content. The book is intended for physicists, astrophysicists, cosmologists, mathematicians and astronomers.

It is a pleasure to thank many colleagues for their interest and comments on the content of this book. I am especially indebted to Yuval Ne'eman for his continuous interest and encouragement, and for his many illuminating remarks. Thanks are also due to Gidon Erez for his constant interest and for the many suggestions and comments which led to a better presentation of the subject, and to Alex Gersten for the drawing of some of the figures in the book. I am particularly grateful to Julia Goldbaum for her constant assistance and the excellent job of typing the book, and to Dr. Kenneth Chan, Science Editor, World Scientific Publishing

Preface ix

Co., for his cooperation. The permission granted by the Albert Einstein Archives, the Hebrew University of Jerusalem, Israel, to quote from Einstein's work is gratefully acknowledged.

Beer Sheva, Israel   *Moshe Carmeli*
January 1997

In the second edition two more chapters have been added: In Chapter 8 the relationship between velocity, acceleration and cosmic distances is given, and Chapter 9 deals with the first days after the Big Bang of the Universe. Also added three appendices that deal with the introduction of gravity to the theory: Appendix A presents Cosmological General Relativity, a four-dimensional space-velocity theory of gravitation. Appendix B presents a five-dimensional unified theory of gravitation that includes space, time and velocity. Finally in Appendix C the important problem of the cosmic temperature decline during the evolution of the Universe, starting with the Big Bang and up to our present time at which the temperature is 2.73K, is given.

I am indebted to Professor Adam Riess for his permission to reproduce some of the figures appearing in this book, and for his suggestion of the cosmic time at which the Universe had a constant expansion. Finally, many thanks are due to Julia Goldbaum for typing the second edition of the book, and to Ms E.H. Chionh, Editor, WS Publishing Co., for her cooperation.

February 2002   *Moshe Carmeli*

Contents

1 Introduction — 1
  1.1 Historical background 1
  1.2 Cosmology and special relativity 2
  1.3 References 5

2 Cosmological Special Relativity — 7
  2.1 Introduction 8
  2.2 Fundamentals of special relativity 8
  2.3 Present-day cosmology 9
  2.4 Postulates 10
  2.5 Cosmic frames 10
  2.6 Spacevelocity in cosmology 11
  2.7 Pre-special-relativity 11
  2.8 Relative cosmic time 12
  2.9 Inadequacy of the classical transformation 13
  2.10 Universe expansion versus light propagation . . . 13
  2.11 The cosmological transformation 15
  2.12 Interpretation of the cosmological transformation 17
  2.13 Another derivation of the
       cosmological transformation 17
  2.14 The galaxy cone 18
  2.15 Consequences of the cosmological transformation 20
       2.15.1 Classical limit 20

	2.15.2 Length contraction	20
	2.15.3 Velocity contraction	22
	2.15.4 Law of addition of cosmic times	23
	2.15.5 Inflation of the Universe	24
	2.15.6 Minimal acceleration in nature	25
	2.15.7 Cosmological redshift	26
2.16	Concluding remarks	26
2.17	References	27

3 Extension of the Lorentz Group to Cosmology   29
3.1	Preliminaries	29
3.2	The line element	32
3.3	The transformations explicitly	32
3.4	The generalized transformation	34
3.5	Concluding remarks	35
3.6	References	36

4 Fundamentals of Einstein's Special Relativity   37
4.1	Postulates of special relativity	38
	4.1.1 The principle of relativity. Constancy of the speed of light	38
	4.1.2 Coordinates	39
	4.1.3 Inertial coordinate system	39
	4.1.4 Simultaneity	40
4.2	The Galilean transformation	40
	4.2.1 The Galilean group	41
4.3	The Lorentz transformation	41
	4.3.1 Measuring rods and clocks	42
	4.3.2 Spatial coordinates and time	42
	4.3.3 Einstein's paradox	42
	4.3.4 Apparent incompatibility of the special relativity postulates	43
	4.3.5 Remark on action-at-a-distance	44

	4.3.6 Derivation of the Lorentz transformation	44
	4.3.7 The Lorentz group	52
	4.3.8 Problems	54
4.4	Consequences of the Lorentz transformation	59
	4.4.1 Nonrelativistic limit	59
	4.4.2 The Lorentz contraction of lengths	60
	4.4.3 The dilation of time	61
	4.4.4 The addition of velocities law	61
	4.4.5 Problems	63
4.5	References	66

5 Structure of Spacetime · 67
5.1	Special relativity as a valuable guide	68
5.2	Four dimensions in classical mechanics	68
5.3	The Minkowskian spacetime	69
5.4	Proper time	72
5.5	Velocity and acceleration four-vectors	75
5.6	Problems	78
5.7	References	78

6 The Light Cone · 81
6.1	The light cone	81
6.2	Events and coordinate systems	82
6.3	Problems	84
6.4	Future and past	84
6.5	References	85

7 Mass, Energy and Momentum · 87
7.1	Preliminaries	88
7.2	Mass, energy and momentum	88
7.3	Angular-momentum representation	92

7.4	Energy-momentum four-vector	95
7.5	Problems	96
7.6	References	97

8 Velocity, Acceleration and Cosmic Distances — 99

8.1	Preliminaries	99
8.2	Velocity and acceleration four-vectors	100
8.3	Acceleration and distances	102
8.4	Energy in ESR versus cosmic distance in CSR	104
8.5	Distance-velocity four-vector	104
8.6	Conclusions	106
8.7	References	107

9 First Days of the Universe — 109

9.1	Preliminaries	109
9.2	Lengths of days	110
9.3	Comparison with Einstein's special relativity	111
9.4	References	112

A Cosmological General Relativity — 115

A.1	Preliminaries	116
A.2	Cosmology in spacevelocity	117
A.3	Gravitational field equations	120
A.4	Solution of the field equations	123
A.5	Classification of universes	125
A.6	Physical meaning	127
A.7	The accelerating universe	128
A.8	Theory versus experiment	136
A.9	Concluding remarks	139
A.10	References	143

B Five-Dimensional Brane World Theory — 145

- B.1 Introduction 146
  - B.1.1 Cosmic coordinate systems: The Hubble transformation 146
  - B.1.2 Lorentz-like cosmological transformation 148
  - B.1.3 Five-dimensional manifold of space, time and velocity 149
- B.2 Universe with gravitation 150
  - B.2.1 The Bianchi identities 151
  - B.2.2 The gravitational field equations 151
  - B.2.3 Velocity as an independent coordinate . . 152
  - B.2.4 Effective mass density in cosmology 153
- B.3 The accelerating Universe 154
  - B.3.1 Preliminaries 154
  - B.3.2 Expanding Universe 156
  - B.3.3 Decelerating, constant and accelerating expansions 158
  - B.3.4 Accelerating Universe 159
- B.4 The Tully-Fisher formula: Halo dark matter . . . 160
  - B.4.1 The geodesic equation 161
  - B.4.2 Equations of motion 163
  - B.4.3 The Tully-Fisher law 166
- B.5 The cosmological constant 168
  - B.5.1 The cosmological term 168
  - B.5.2 The supernovae experiments value for the cosmological constant 170
  - B.5.3 The Behar-Carmeli predicted value for the cosmological constant 170
  - B.5.4 Comparison with experiment 171
- B.6 Cosmological redshift analysis 172
  - B.6.1 The redshift formula 172
  - B.6.2 Particular cases 173

	B.6.3 Conclusions	175
B.7	Concluding remarks	175
B.8	Mathematical conventions and Christoffel symbols	176
B.9	Components of the Ricci tensor	177
B.10	Integration of the Universe expansion equation	178
B.11	References	180
C	**Cosmic Temperature Decline**	**185**
C.1	Introduction	185
C.2	Temperature formula without gravity	186
C.3	Comparison	187
C.4	References	188

Index   **189**

Chapter 1

Introduction

1.1 Historical background

In 1905 A. Einstein [1] published his paper on the electrodynamics of moving charges, later on to be known as the theory of special relativity. The main objective of the theory was to establish the compatibility of the constancy of the propagation of light with the (apparently contradictory) validity of all physical laws (and particularly the constancy of the speed of light in vacuum) in all inertial frames. His conclusion was that the Galilean transformation should be abandoned in favor of the Lorentz transformation which he rederived and gave to it the right physical meaning.

The absolute-space concept was consequently shown to be wrong. So was the absolute time. Only their combination as one entity as spacetime, as was later on formulated by H. Minkowski [2], has a physical meaning.

Realizing that gravitation cannot be accommodated within

the framework of special relativity theory, and that he should go to curved spacetime, Einstein developed the theory of general relativity.

Years later, the Russian mathematician A. Friedmann [3] solved Einstein's gravitational field equations of general relativity and found that they have nonstatic cosmological solutions which indicate on an expanding universe. Einstein, believing that the Universe should be static and unchanged forever, suggested a modification to his field equations by adding to them the so-called cosmological term which cancels the expansion.

Soon after that E. Hubble [4] found experimentally that the far-away galaxies are receding from us, and that the farther the galaxy the bigger its receding velocity. In simple words, the Universe is indeed expanding according to a simple physical law nowadays known as the Hubble law. This brought Einstein to say that the introduction of the cosmological term to his equations was the biggest blunder of his life.

1.2 Cosmology and special relativity

It thus seems that not only light propagates in a constant velocity in vacuum but the Universe also expands with a constant fashion whose proportionality constant is the Hubble constant. But there is an essential difference between them: light propagation is expressed in terms of space and time, whereas the Universe expansion is by space and velocity at a fixed and every specific cosmic time. Thus we are dealing with what might be called the "dual" space in cosmology as compared to the ordinary space in light propagation. One can thus deal with cosmology as is the situation in thermodynamics, namely working with a theory whose dynamical variables are also the observable measured quantities. This observation will be expounded and utilized throughout this monograph.

The large-scale structure of the cosmos is analysed in details in this monograph under the assumption that gravitation is *negligible* and thus the space is flat. A comparison is made between the present-day cosmology and the prerelativistic physics; it is shown that there is an analogy between the structure of space and time as it is understood these days, and the structure of space and velocity in cosmology. The cosmic time takes the role of the velocity; observers at different locations in the Universe are related to each other by *relative* (rather than absolute) cosmic times, just as observers in different inertial systems are related to each other by relative velocities. As a result, we are able to develop a special relativity theory of cosmology which unites space and velocity, in complete analogy to Einstein's special relativity theory that unites space and time.

Chapter 2 gives the detailed consequences of the Hubble law at each instant of cosmic time. Accordingly, the cosmological four-dimensional transformation, that relates the spatial coordinates to the outgoing radial velocity at two different cosmic times, is derived. This is like the Lorentz transformation, but now it is in the dual space of distance and velocity. Several important physical consequences are drawn, among which is the inflationary universe at the very early cosmic time. At the very early stage of the Universe gravitation is negligible and the transformation obtained can therefore be used. It thus appears that the inflation of space at the early stage of the Universe is a property of spacetime and one does not need a specific model of material to cause the inflation. Likewise, a cosmological-relativistic law of addition of cosmic times is derived.

In Chapter 3 it is shown how space, time and the velocity can be combined into a higher-dimensional space, and the relevant transformation, an extension of the Lorentz transformation, is derived. The group of the transformations is presented. The extended group includes as subgroups the homogeneous Lorentz group (preserving the constancy of the velocity of light in vacuum), the cosmological group (preserving the constancy of the rate of expansion of the Universe at each cosmic time), and the four-dimensional rotation group of cosmic time and velocities (that preserves the distances).

Chapters 4–7 give an extended exposition of special relativity theory, which is presented along the original lines of Einstein rather than dealing with technicalities. Chapter 4 starts with the postulates of the theory, the principle of relativity and the principle of the constancy of the speed of light in vacuum. Subsequently, the basic concepts of coordinate systems, and particularly the inertial systems, are discussed. The essential notion of simultaneity is analysed. These concepts are subsequently followed by the Galilean transformation and its group, and the Lorentz transformation and its group. Consequences of the Lorentz transformation are drawn, including the familiar contraction of length, the dilation of time, and the relativistic formula of addition of velocities.

Following Minkowski, a four-dimensional formulation of space and time is given in Chapter 5. This includes the concepts of proper time, four-vectors, and the velocity and acceleration four-vectors. The light cone structure is then given in Chapter 6. In Chapter 7, the relationship among the mass, energy and momentum, along with the energy-momentum four-vector, is given. The latter four chapters are written in a self-contained fashion and may be used as a text for the theory of special relativity. In Chapter 8 the relationships between cosmic velocity, acceleration and distances are given, and in Chapter 9 the days after the Big Bang are discussed. In Appendix A gravitation is added and, as a result, we have the cosmological general relativity theory. In Appendix B we have a five-dimensional unified theory of space, time and velocity, and in Appendix C the effect of gravity on the cosmic temperature is discussed.

1.3 References

[1] A. Einstein, *Ann. Physik* (Germany) **17**, 891 (1905); English translation in: A. Einstein *et al.: The Principle of Relativity* (Dover, New York, 1923).
[2] H. Minkowski, Space and time (an address delivered at the 80th Assembly of German Natural Scientists and Physicians, at Cologne, Germany, 21 September, 1908); English translation in: *The Principle of Relativity* (Dover, New York, 1923), p. 73.
[3] A. Friedmann, *Z. Phys.* **10**, 377 (1922).
[4] E. Hubble, *The Realm of the Nebulae* (Yale University Press, New Haven, 1936); reprinted by Dover Publications, Inc., New York, 1958.

Chapter 2

Cosmological Special Relativity

In this chapter we present the cosmological special relativity theory along the lines of Einstein's special relativity. The chapter starts by giving the foundations of ordinary special relativity. This is then followed by reviewing the present-day status of cosmology. The postulates of the theory are given and the notion of cosmic frame is introduced. Spacevelocity and relative cosmic time are subsequently discussed. The inadequacy of the classical transformation in physics is discussed and a comparison of the Universe expansion to the light propagation is given. The transformation between spacevelocity coordinates at different cosmic times is derived and its physical interpretation is given. Another derivation of the transformation is presented. Consequences of the transformation are then drawn. The chapter ends with the concluding remarks.

2.1 Introduction

In prerelativistic physics it was assumed that space is not related to time; a "stationary" frame of reference was presumed to exist with respect to which all physical phenomena can be described.

As Einstein [1] showed, in his famous paper of 1905, this picture was wrong; space has no preference to a particular frame on any other one that moves with a constant velocity, and in this way one can accommodate the fact that light propagates with a constant velocity in all moving systems. The mixture of space and time became a necessity in order to preserve the constancy of the propagation of light in all inertial frames. The mathematical expression of this fact is given by the familiar Lorentz transformation which was rederived by Einstein who also gave it the correct physical interpretation.

As Bernard Russell said, "Einstein's theory of relativity is probably the greatest synthetic achievement of the human intellect up to the present time."

2.2 Fundamentals of special relativity

The essence of the theory of special relativity, as Einstein put it in his Autobiographical Notes, is as follows [2].

"According to the rules of connection, used in classical physics, between the spatial coordinates and the time of events in the transition from one inertial system to another, the two assumptions of

(1) the constancy of the light velocity
(2) the independence of the laws (thus especially also of the law of the constancy of the light velocity) from the choice of inertial system (principle of special relativity)

are mutually incompatible (despite the fact that both taken separately are based on experience)".

"The insight fundamental for the special theory of relativity is this: The assumptions (1) and (2) are compatible if relations of a new type (Lorentz transformation) are postulated for the conversion of coordinates and times of events. With the given physical interpretation of coordinates and time, this is by no means merely a conventional step but implies certain hypotheses concerning the actual behavior of moving measuring rods and clocks, which can be experimentally confirmed or disproved."

"The universal principle of the special theory of relativity is contained in the postulate: The laws of physics are invariant with respect to Lorentz transformations (for the transition from one inertial system to any other arbitrarily chosen inertial system). This is a restricting principle for natural laws, comparable to the restricting principle of the nonexistence of the *perpetuum mobile* that underlies thermodynamics."

2.3 Present-day cosmology

We wish to point out that at present we have a similar situation in cosmology to that existed in prerelativistic times with respect to space and (not velocity but) cosmic time, in conjunction with the constancy of expansion of the Universe (and not propagation of light). If we make the convention according to which cosmic time, denoted by t, is measured *backward*, then our present time ($t = 0$) is a preferred time with respect to which all cosmological physical phenomena are referred. This is exactly analogous to the prerelativity assumption that physical phenomena are referred to only one "stationary" ($v = 0$) system [3,4].

Actually space has no such a preference: When we consider an astronomical object and say that it is, let us say, at $t = \tau/2$, where $\tau = 1/H_0$ is Hubble's time, that faraway object has the same right to say that he is at cosmic time zero ($t = 0$) and we are at $t = \tau/2$ with respect to him, exactly as in relativistic physics but with the roles of cosmic time and velocity exchanged. We will assume that such a reciprocity relationship between cosmological objects is a universal property of space and cosmic time just as Einstein did with respect to space and velocity in special relativity.

2.4 Postulates

In addition, we will make two assumptions which will be elevated to postulates. These are: (1) The *principle of the constancy of the expansion of the Universe* (expressed by Hubble's law) at all cosmic times (analogous to the principle of the constancy of propagation of light in all moving frames); and (2) The *principle of cosmological relativity* (analogous to the principle of special relativity) according to which the laws of physics are the same at all cosmic times (as moving frames in special relativity).

2.5 Cosmic frames

In this way the Universe has *cosmic frames of reference* located at fixed cosmic times and differ from each other by relative constant cosmic times, similar to the situation in special relativity but now cosmic times replace velocities. Observers in each cosmic frame are equipped with rulers to measure distances (like in special relativity) and with small radar devices (similar to those used by highway patrol) for velocity measurements (instead of clocks in special relativity). Notice the analogy between the relation $[\tau]$ =distance/velocity in the present theory and $[c]$ =distance/time in special relativity, which suggests the choice of distance and velocity as our fundamental variables as compared to distance and time in special relativity.

Remark: The constant τ is used by us just as the constant c is used in special relativity, even though it is well known that both the speed of light and the rate of expansion of the Universe change their values due to gravity. This is possible since local measurements of both the velocity of light and the rate of expansion of the Universe always yield constant c and τ, respectively.

2.6 Spacevelocity in cosmology

With the above postulates, and by comparison with special relativity, it is obvious that space and velocity cannot be independent if Hubble's law is to be preserved at all cosmic times. In fact this will enable us to derive a transformation that relates space points and velocities (and other quantities) measured in different cosmic frames of reference that differ in relative cosmic times just like the Lorentz transformation which relates space points and times (and other quantities) measured in different inertial frames that differ in relative velocities. Space coordinates and velocities become unified in cosmology just as space and time are unified in local (noncosmological) physics.

2.7 Pre-special-relativity

With the above preliminaries we are now in a position to develop the theory. To begin with we repeat very briefly what preceded to special relativity. The Galilean transformation between two inertial systems K and K', where K' moves relative to K with a constant velocity v along the x axis, is given by

$$x' = x - vt, \quad t' = t, \quad y' = y, \quad z' = z.$$

Here x and x' represent the coordinates of a particle in the systems K and K', respectively.

The problem with the Galilean transformation is its incompatibility with the equation of propagation of light which satisfies

$$c^2 t'^2 - x'^2 = c^2 t^2 - x^2, \quad y' = y, \quad z' = z.$$

Hence the Galilean transformation should be abolished in favor of a new one that relates not only x' to x leaving t unchanged but relates x' and t' to x and t. And this immediately leads to the familiar Lorentz transformation.

2.8 Relative cosmic time

In cosmology one is not interested in comparing quantities at two reference frames moving with a constant velocity with respect to each other. Rather, one is interested in comparing quantities at two different cosmic times. For example, one often asks what was the density of matter or the temperature of the Universe at an earlier cosmic time t as compared to the values of these quantities at our present time now ($t = 0$). The backward time t is the *relative* cosmic time with respect to our present time.

The concept of the relative cosmic time is not restricted only to the backward cosmic time t with respect to the present time ($t = 0$). Every two observers with cosmic times t_1 and t_2 with respect to us are related to each other by a relative cosmic time t. Thus t plays the role of the velocity v in special relativity and we will see in the sequel that t has an upper limit which is the Hubble time τ just as the maximum velocity permitted in special relativity is c.

The variables (coordinates) in this theory are naturally the Hubble variables, i.e. the velocity v and the distance x. To derive the transformation between these variables in the systems K and K', where K' has a relative cosmic time t with respect to K, we proceed as follows.

2.9 Inadequacy of the classical transformation

We first do it classically, and for simplicity it is assumed that the motion is one-dimensional. Denoting the coordinates and velocities in the systems K and K' by x, v and x', v', respectively, then
$$x' = x - tv, \quad v' = v, \quad y' = y, \quad z' = z,$$
where v was assumed to be constant. The x's and v's in these equations represent the coordinates and velocities not for just one particle but for as many as one wishes, with t the same for all of them.

The above transformation does not satisfy the equation of expansion of the Universe which, according to the principle of the constancy of expansion of the Universe and the principle of cosmological relativity demanding the laws of physics (and in particular Hubble's law) to be valid at all cosmic times, satisfies
$$\tau^2 v'^2 - x'^2 = \tau^2 v^2 - x^2, \quad y' = y, \quad z' = z.$$
The situation here is similar to what we had at the beginning of the century where the Galilean transformation could not accommodate both the principle of special relativity and the principle of constancy of the speed of light, whence leading to the Lorentz transformation. A new transformation here also has to be found, which relates not only x' to x leaving v unchanged but relates x' and v' to x and v.

2.10 Universe expansion versus light propagation

Under the assumption that Hubble's constant is constant in cosmic time, there is an analogy between the propagation of light,

$x = ct$, and the expansion of the Universe, $x = \tau v$, where τ is Hubble's time, a constant which is also the age of the Universe under the above assumption, and c is the speed of light in vacuum. Thus one can express the expansion of the Universe, assuming that it is homogeneous and isotropic, in terms of the null vector (v, x, y, z) satisfying

$$\tau^2 v^2 - \left(x^2 + y^2 + z^2\right) = 0, \qquad (2.1)$$

where v is the receding velocity of the galaxies. Equation (2.1), in the 4-dimensional flat space of the Cartesian 3-space and the velocity, is similar to

$$c^2 t^2 - \left(x^2 + y^2 + z^2\right) = 0 \qquad (2.2)$$

for the null propagation of light in Minkowskian spacetime. We assume, furthermore, that a relationship of the form (2.1) is valid at all cosmic times. Thus, at a cosmic time t' at which the coordinates and velocity are labeled with primes, we have

$$\tau^2 v'^2 - \left(x'^2 + y'^2 + z'^2\right) = 0 \qquad (2.3)$$

with the same τ, just as for light emitted from a source with velocity v with respect to the first one,

$$c^2 t'^2 - \left(x'^2 + y'^2 + z'^2\right) = 0. \qquad (2.4)$$

Accordingly, we have a 4-dimensional space with zero curvature of x, y, z, v just as the Minkowskian spacetime of x, y, z, t.

We now assume that at two cosmic times t and t' we have

$$\tau^2 v'^2 - \left(x'^2 + y'^2 + z'^2\right) = \tau^2 v^2 - \left(x^2 + y^2 + z^2\right), \qquad (2.5)$$

in analogy to the special relativistic formula

$$c^2 t'^2 - \left(x'^2 + y'^2 + z'^2\right) = c^2 t^2 - \left(x^2 + y^2 + z^2\right). \qquad (2.6)$$

The question is then what is the transformation between x', y', z', v' and x, y, z, v that satisfies the invariance formula (2.5).

2.11 The cosmological transformation

For simplicity we again assume that the motion is along the x axis. Hence Hubble's law in the systems K and K' is given by

$$x = \tau v, \quad x' = \tau v', \tag{2.7}$$

where x, v and x', v' are measured in K and K'. Assuming now that x, v and x', v' transform linearly, then

$$x' = ax - bv, \tag{2.8}$$

$$x = ax' + bv', \tag{2.9}$$

where a and b are some variables which are independent of the coordinates.

At $x' = 0$ and $x = 0$, Eqs. (2.8) and (2.9) yield, respectively,

$$\frac{b}{a} = \frac{x}{v} = t, \tag{2.10}$$

and

$$\frac{b}{a} = -\frac{x'}{v'} = t. \tag{2.11}$$

Using now Eqs. (2.7), (2.8) and (2.9) we obtain

$$\tau v = x = ax' + bv' = a\tau v' + bv' = (a\tau + b) v', \tag{2.12a}$$

and similarly

$$\tau v' = (a\tau - b) v. \tag{2.12b}$$

Eliminating v and v' from Eqs. (2.12), and using $b = at$ from Eq. (2.10), we get

$$\tau^2 = a^2\left(\tau^2 - t^2\right), \tag{2.13}$$

or

$$a = \frac{1}{\sqrt{1 - \frac{t^2}{\tau^2}}}, \tag{2.14}$$

and therefore

$$b = \frac{t}{\sqrt{1 - \frac{t^2}{\tau^2}}}. \tag{2.15}$$

Inserting these results in Eqs. (2.8) and (2.9) we obtain

$$x' = \frac{x - tv}{\sqrt{1 - \frac{t^2}{\tau^2}}}, \tag{2.16a}$$

$$v' = \frac{v - xt/\tau^2}{\sqrt{1 - \frac{t^2}{\tau^2}}}, \tag{2.16b}$$

$$y' = y, \quad z' = z, \tag{2.16c}$$

and

$$x = \frac{x' + tv'}{\sqrt{1 - \frac{t^2}{\tau^2}}}, \tag{2.17a}$$

$$v = \frac{v' + x't/\tau^2}{\sqrt{1 - \frac{t^2}{\tau^2}}}, \tag{2.17b}$$

$$y = y', \; z = z', \tag{2.17c}$$

for the inverse transformation. Equations (2.16) and (2.17) will be referred to as the *cosmological transformation*.

2.12 Interpretation of the cosmological transformation

Equations (2.16) give the transformed values of x and v as measured in the system K' with a relative cosmic time t with respect to K. The roles of the time and the velocity are *exchanged* as compared to special relativity. This fits our needs in cosmology where one measures distances and velocities at different cosmic times in the past. The parameter t/τ replaces v/c of special relativity.

It should be emphasized that the transformation (2.16) is not a trivial exchange of v/c, appearing in the Lorentz transformation, and t/τ here. For example, the redshift $z = v/c$ at low velocities, but is certainly not equal to t/τ for small t/τ. (Details on the redshift are given in the sequel.)

2.13 Another derivation of the cosmological transformation

The transformation (2.16) could also have been derived like deriving the Lorentz transformation in standard texts (see Chapter 4) by writing

$$x'^2 - \tau^2 v'^2 = x^2 - \tau^2 v^2, \tag{2.18}$$

whose solution is

$$x' = x \cosh \psi - \tau v \sinh \psi,$$
$$\tau v' = \tau v \cosh \psi - x \sinh \psi. \qquad (2.19)$$

At $x' = 0$ we obtain

$$\tanh \psi = \frac{x}{\tau v} = \frac{t}{\tau}, \qquad (2.20)$$

and therefore

$$\sinh \psi = \frac{t/\tau}{\sqrt{1 - \frac{t^2}{\tau^2}}}, \qquad (2.21a)$$

$$\cosh \psi = \frac{1}{\sqrt{1 - \frac{t^2}{\tau^2}}}, \qquad (2.21b)$$

which lead to the transformation (2.16).

2.14 The galaxy cone

The invariant equation (2.1), describing the distribution of galaxies in the Universe at any cosmic time, has a very simple geometrical interpretation. It enables one to present the locations of galaxies as a cone in the dual space of distance and velocity. One then has a *galaxy cone*, similar to the familiar light cone in special relativity (see Figure 4.1 of Chapter 4). The symmetry axis of the cone coincides with the x^0 axis which extends from $-\tau c$ to $+\tau c$.

The galaxy cone

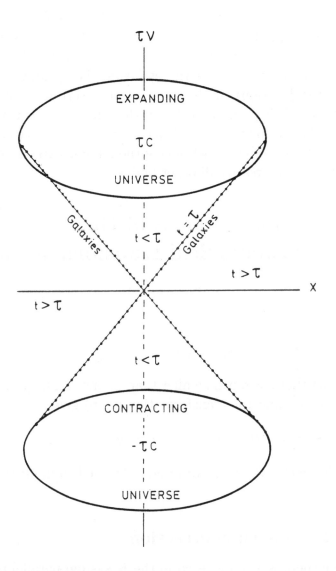

Figure 2.1 The galaxy cone in cosmological relativity, describing the cone in the $x - v$ space satisfying $x^2 - \tau^2 v^2 = 0$, where x represents the three-dimensional space. The heavy dots describe galaxies. The galaxy cone represents the locations of the galaxies at a given time rather than their path of motion in the real space.

Figure 2.1 describes the galaxy cone in cosmological relativity. It gives the description of the cone in the $x - v$ space satisfying $x^2 - \tau^2 v^2 = 0$, where x stands for the three-dimensional space. The heavy dots describe galaxies. The galaxy cone represents the locations of the galaxies at a certain cosmic time rather than their path of motion in the real space (as is the case for light in the light cone). While points at the surface of the cone represent bodies which follow the Hubble expansion, those in its interior represent all other bodies.

2.15 Consequences of the cosmological transformation

In the following we draw some consequences of the cosmological transformations (2.16) and (2.17).

2.15.1 Classical limit

Assuming that t is much smaller than τ, one can neglect t^2 with respect to τ^2, and the transformation (2.16) gives

$$x' = x - tv, \quad v' = v, \quad y' = y, \quad z' = z, \qquad (2.22)$$

which is exactly the transformation obtained from classical mechanics.

2.15.2 Length contraction

Suppose there is a rod located in the K system parallel to the x axis. Let its length, measured in this system, be $\Delta x = x_2 - x_1$, where x_1 and x_2 are the coordinates of the two ends of the rod. To determine the length of this rod as measured in the K' system we must find the coordinates of the two ends of the rod x'_1 and

Consequences of the cosmological transformation

x_2' in this system at the same velocity v'. From Eqs. (2.17) we have

$$x_1 = \frac{x_1' + tv'}{\sqrt{1 - \frac{t^2}{\tau^2}}},$$

$$x_2 = \frac{x_2' + tv'}{\sqrt{1 - \frac{t^2}{\tau^2}}}.$$

The length of the rod in the K' system is $\Delta x' = x_2' - x_1'$, thus

$$\Delta x = \frac{\Delta x'}{\sqrt{1 - \frac{t^2}{\tau^2}}}.$$

The *proper* length of a rod is its length in a system in which it is located. Let us denote it by $L_0 = \Delta x$ and the length of the rod in any other system K' by L. Then

$$L = L_0 \sqrt{1 - \frac{t^2}{\tau^2}}. \qquad (2.23)$$

Thus a rod has its greatest length in the system in which its relative cosmic time with respect to the system is zero; its length in a system in which it is located at a relative cosmic time t with respect to that system is decreased by the factor $(1 - t^2/\tau^2)^{1/2}$. This result of the present theory is exactly similar to the familiar Lorentz contraction with the factor $(1 - v^2/c^2)^{1/2}$ in special relativity given in Subsection 4.4.2.

2.15.3 Velocity contraction

Suppose a velocity measuring instrument is located at $x' = 0$ in the K' system. Then from Eqs. (2.17) we have

$$v = \frac{v'}{\sqrt{1 - \frac{t^2}{\tau^2}}}. \tag{2.24}$$

Denoting now v by v_0 and v' by v we obtain

$$v = v_0 \sqrt{1 - \frac{t^2}{\tau^2}}. \tag{2.25}$$

The above result is like the time dilation in special relativity (see Subsection 4.4.3) and was expected since time in special relativity goes over to the velocity in the present theory. The velocity measured by an observer with a relative cosmic time t with respect to us is smaller by the factor $(1 - t^2/\tau^2)^{1/2}$ than what is observed by us at $t = 0$.

Remark on Dark Matter: As is well known much of the support for the existence of the dark matter is due to the observed very high velocities of gas molecules or galaxies. For example galaxies in the far-off Coma cluster are observed whirling around one another faster than the laws of physics would allow. So is the mysteriously rapid rotation of spiral galaxies. Equation (2.25) clearly shows that the observed velocity by us is not the velocity measured by a local observer at a relative time t with respect to us. He measures a smaller velocity, and the more back in time the more the velocity decreases. Does this mean that the hypothetical dark matter can be abolished just as the "luminiferous ether" was proved to be superfluous by special relativity?

2.15.4 Law of addition of cosmic times

Dividing the first of Eqs. (2.17) by the second we find, choosing $t = t_1$,

$$\frac{x}{v} = \frac{x' + t_1 v'}{v' + \frac{t_1}{\tau^2} x'}, \qquad (2.26)$$

or, dividing the numerator and the denominator of the right-hand side of this equation by v', we obtain

$$t = \frac{t_1 + t_2}{1 + \frac{t_1 t_2}{\tau^2}}, \qquad (2.27)$$

where $t_2 = x'/v'$ and $t = x/v$.

Equation (2.27) determines the transformation of cosmic time and describes the law of composition of cosmic times in the Universe. In the limiting case of t much smaller than the Hubble time τ, Eq. (2.27) goes over to the formula $t = t_1 + t_2$ of classical physics.

We see that the simple law of adding and subtracting cosmic times is no longer valid or, more precisely, is only approximately valid for short times with respect to us, but not for those near the Hubble time, which is also the age of the Universe in this case. Two consecutive events that occur at $t_1 = (9/10)\tau$ and $t = (180/181)\tau$ both with respect to us (at $t = 0$), for example, then with respect to the first event the second one does not occur at $t - t_1 \approx \tau/10$ but rather at

$$t_2 = \frac{t - t_1}{1 - \frac{t t_1}{\tau^2}} = \frac{9}{10}\tau, \qquad (2.28)$$

which is much longer than $t - t_1$ and happens to be exactly equal to t_1. We also notice that the past cosmic time cannot be greater than τ, the age of the Universe. This is similar to what we have in special relativity where the velocity cannot exceed c (see Subsection 4.4.4). It will be noted that one may add as many successive time intervals as one wishes without ever reaching the age of the Universe τ.

2.15.5 Inflation of the Universe

The line element is given by

$$\tau^2 dv^2 - \left(dx^2 + dy^2 + dz^2\right) = ds^2. \tag{2.29}$$

Hence

$$\tau^2 \left(\frac{dv}{ds}\right)^2 - \left[\left(\frac{dx}{dv}\right)^2 + \left(\frac{dy}{dv}\right)^2 + \left(\frac{dz}{dv}\right)^2\right] \left(\frac{dv}{ds}\right)^2$$

$$= \left(\tau^2 - t^2\right) \left(\frac{dv}{ds}\right)^2 = 1. \tag{2.30}$$

Multiplying now this equation by ρ_0^2, the matter density of the Universe at the present time, we obtain for the matter density at a past time t

$$\rho = \tau \rho_0 \frac{dv}{ds} = \frac{\rho_0}{\sqrt{1 - \frac{t^2}{\tau^2}}}. \tag{2.31}$$

Since the volume of the Universe is inversely proportional to its density, it follows that the ratio of the volumes at two backward cosmic times t_1 and t_2 with respect to us is given by $(t_2 < t_1)$.

$$\frac{V_2}{V_1} = \sqrt{\frac{1 - t_2^2/\tau^2}{1 - t_1^2/\tau^2}} = \sqrt{\frac{(\tau - t_2)(\tau + t_2)}{(\tau - t_1)(\tau + t_1)}}. \quad (2.32)$$

For times t_1 and t_2 very close to τ we can assume that $\tau + t_2 \approx \tau + t_1 \approx 2\tau$. Hence

$$\frac{V_2}{V_1} = \sqrt{\frac{T_2}{T_1}}, \quad (2.33)$$

where $T_1 = \tau - t_1$ and $T_2 = \tau - t_2$. For $T_2 - T_1 \approx 10^{-32}$ sec and $T_2 \ll 1$ sec, we then have

$$\frac{V_2}{V_1} \approx \sqrt{1 + \frac{10^{-32}}{T_1}} \approx \sqrt{\frac{10^{-32}}{T_1}} = \frac{10^{-16}}{\sqrt{T_1}}. \quad (2.34)$$

For $T_1 \approx 10^{-132}$ sec we obtain $V_2 \approx 10^{50} V_1$.

The above result conforms with inflationary universe theory without assuming any model (such as the Universe is propelled by a sort of antigravity) [5,6].

2.15.6 Minimal acceleration in nature

From Eq. (2.10) we have

$$t = \frac{x}{v} = \frac{dx}{dv} = \frac{v}{a}, \quad (2.35)$$

where a is the acceleration. Hence

$$t_{max} = \tau = \left(\frac{v}{a}\right)_{max} = \frac{c}{a_{min}}. \quad (2.36)$$

It thus appears that in nature there is a minimal acceleration

$$a_{min} = \frac{c}{\tau} \approx 10^{-8} \text{cm/sec}^2. \tag{2.37}$$

It will be noted that such a minimal acceleration constant has been proposed by Milgrom to exist on intuitive basis, but without explaining its origin [7].

2.15.7 Cosmological redshift

The wave length of light is inversely proportional to the interval of length as measured by two observers at different cosmic times. The result is

$$\frac{\lambda}{\lambda_0} = \left(1 - \frac{t^2}{\tau^2}\right)^{-1/2}. \tag{2.38}$$

Thus the wave length of light emitted from sources back in time increases as compared to its value on Earth. For $t/\tau \ll 1$, we have

$$z = \frac{\lambda}{\lambda_0} - 1 \approx \frac{1}{2}\frac{t^2}{\tau^2}. \tag{2.39}$$

2.16 Concluding remarks

The above cosmological special relativity corresponds to a Universe with zero curvature, i.e. $\Omega = \rho_0/\rho_c = 1$, thus $\rho_0 = \rho_c = 3/8\pi G\tau^2 \approx 10^{-29}\text{g/cm}^3$, a few hydrogen atoms per cubic meter, is the vacuum energy density, and ρ_0 is the present-time mean mass density. Due to the flatness of the spacevelocity in this particular case, and only in this case (other cases are $\Omega > 1$ and $\Omega < 1$), a cosmological special relativity could have been developed since in the $\Omega > 1$ and $\Omega < 1$ cases the spacevelocity is not flat.

In a sense the theory presented here is half dynamical since $\rho_0 \neq 0$ as opposed to ordinary special relativity which is in a sense can be considered as kinematical. It is for this reason that we could obtain results similar to those obtained from the inflationary universe model.

2.17 References

[1] A. Einstein, *Ann. Physik* (Germany) **17**, 891 (1905); English translation in: A. Einstein *et al.: The Principle of Relativity* (Dover, New York, 1923).
[2] A. Einstein, Autobiographical Notes, translated and edited by P.A. Schilpp (Open Court Publishing Co., La Salle and Chicago, 1979).
[3] M. Carmeli, *Found. Phys.* **25**, 1029 (1995); **26**, 413 (1996).
[4] M. Carmeli, *International J. Theor. Phys.* **36**, March 1997.
[5] A.H. Guth, *Phys. Rev. D* **23**, 347 (1982).
[6] A.D. Linde, *Phys. Letters* **116B**, 335 (1982).
[7] M. Milgrom, *Astrophys. J.* **270**, 365, 371 and 384 (1983).

Chapter 3

Extension of the Lorentz Group to Cosmology

In this chapter an extension of the familiar Lorentz group to cosmology is presented. The extended group includes as subgroups the homogeneous Lorentz group (preserving the constancy of the velocity of light in vacuum), the cosmological group (preserving the constancy of the rate of expansion of the Universe at each cosmic time) and a four-dimensional rotation group of transformations of time and three-velocities (which preserves distances).

3.1 Preliminaries

In this chapter we present the group of transformations which includes as subgroups the Lorentz group (preserving the constancy of the rate of propagation of light) and the cosmological group (preserving the constancy of the expansion rate of the Universe - Hubble's law). It follows that there is another subgroup of four-dimensional transformations of the time and the three-velocities

Chapter 3. Extension of the Lorentz Group to Cosmology

which preserves distances [1,2].

The extended group is given by the aggregate of all linear transformations of the seven "coordinates"

$$x^\mu = (ct,\ x,\ y,\ z,\ \tau v_x,\ \tau v_y,\ \tau v_z) \qquad (3.1)$$

satisfying

$$c^2 t'^2 - \left(x'^2 + y'^2 + z'^2\right) + \tau^2 \left(v'^2_x + v'^2_y + v'^2_z\right)$$
$$= c^2 t^2 - \left(x^2 + y^2 + z^2\right) + \tau^2 \left(v^2_x + v^2_y + v^2_z\right), \qquad (3.2)$$

where $\tau = H_0^{-1}$ is the Hubble time, H_0 is Hubble's constant and c is the speed of light in vacuum.

A subtransformation of the above group is obtained if one takes the v's to be unchanged. We then have the four-dimensional Lorentz transformation

$$(t,\ x,\ y,\ z) \to (t',\ x',\ y',\ z') \qquad (3.3)$$

with

$$c^2 t'^2 - \left(x'^2 + y'^2 + z'^2\right) = c^2 t^2 - \left(x^2 + y^2 + z^2\right). \qquad (3.4)$$

This is suitable for observers located at *inertial* frames moving with constant speeds. The parameter of the transformation between the coordinates of two frames is V/c, where V is the relative speed between them.

For t = invariant, Eq. (3.2) yields the cosmological transformation

$$(x,\ y,\ z,\ v_x,\ v_y,\ v_z) \to \left(x',\ y',\ z',\ v'_x,\ v'_y,\ v'_z\right) \qquad (3.5)$$

Preliminaries

satisfying

$$\left(x'^2 + y'^2 + z'^2\right) - \tau^2 \left(v_x'^2 + v_y'^2 + v_z'^2\right)$$
$$= \left(x^2 + y^2 + z^2\right) - \tau^2 \left(v_x^2 + v_y^2 + v_z^2\right). \tag{3.6}$$

An observer here, located in a *cosmic* frame, makes observations at a fixed time. The parameter of the transformation between two cosmic frames is now T/τ, where T is the relative cosmic time between them.

And for unchanged x's, Eq. (3.2) yields a four-dimensional rotation

$$(t,\ v_x,\ v_y,\ v_z) \rightarrow \left(t',\ v_x',\ v_y',\ v_z'\right) \tag{3.7}$$

satisfying

$$c^2 t'^2 + \tau^2 \left(v_x'^2 + v_y'^2 + v_z'^2\right) = c^2 t^2 + \tau^2 \left(v_x^2 + v_y^2 + v_z^2\right). \tag{3.8}$$

The frames now are fixed at different points in space. The parameter of the transformation between two frames is X/R, where X is the relative distance between them, and $R = c\tau$.

Each of the above three kinds of transformations provides a group: The Lorentz group O(1,3), the cosmological group O(3,3), and the 4-dimensional rotation group O(4), respectively. The group that includes all of them is O(3,4) as is seen from Eq. (3.2).

3.2 The line element

The line element corresponding to the full group is given by

$$ds^2 = c^2 dt^2 - \left(dx^2 + dy^2 + dz^2\right) + \tau^2 \left(dv_x^2 + dv_y^2 + dv_z^2\right). \tag{3.9}$$

This formula can also be written in a simple form

$$ds^2 = \eta_{\mu\nu} dx^\mu dx^\nu, \tag{3.10}$$

where $\eta_{\mu\nu}$ is a generalized Minkowskian metric in seven dimensions, and whose diagonal is

$$(+1,\ -1,\ -1,\ -1,\ +1,\ +1,\ +1) \tag{3.11}$$

with signature $+1$.

3.3 The transformations explicitly

We give explicitly the transformations described above. With a simplified notation and obvious choice of the coordinates we obtain

$$x' = \frac{x - Vt}{\sqrt{1 - \frac{V^2}{c^2}}}, \tag{3.12a}$$

$$ct' = \frac{ct - Vx/c}{\sqrt{1 - \frac{V^2}{c^2}}} \tag{3.12b}$$

The transformations explicitly

for the Lorentz group (V is the relative velocity),

$$x' = \frac{x - Tv}{\sqrt{1 - \frac{T^2}{\tau^2}}}, \qquad (3.13a)$$

$$\tau v' = \frac{Tv - xT/\tau}{\sqrt{1 - \frac{T^2}{\tau^2}}} \qquad (3.13b)$$

for the cosmological group (T is the relative cosmic time), and

$$ct' = \frac{ct + (X/R)\tau v}{\sqrt{1 + \frac{X^2}{R^2}}}, \qquad (3.14a)$$

$$\tau v' = \frac{\tau v - (X/R)ct}{\sqrt{1 + \frac{X^2}{R^2}}} \qquad (3.14b)$$

for the four-dimensional rotation group (X is the relative distance and $R = c\tau$).

The above discussion may be summarized by stating that one has three kinds of dynamical variables: Distance x, time t, and velocity v along with the following transformations:

$$V/c: \ (x, t) \to (x', t') \ \text{(Lorentz)} \qquad (3.15a)$$

$$T/\tau: \ (x, v) \to (x', v') \ \text{(Cosmological)} \qquad (3.15b)$$

$$X/R: \ (t, v) \to (t', v') \ \text{(Rotation)} \qquad (3.15c)$$

34 Chapter 3. Extension of the Lorentz Group to Cosmology

Thus the highest velocity in nature is c, the maximum time is τ, and the longest distance in the Universe is $R = c\tau$. All of these results essentially follow from the constancy of the speed of light and of Hubble's law.

3.4 The generalized transformation

Using the notation

$$\mathbf{x} = \begin{pmatrix} x \\ ct \\ \tau v \end{pmatrix}, \quad \mathbf{x}' = \begin{pmatrix} x' \\ ct' \\ \tau v' \end{pmatrix}, \qquad (3.16)$$

we can write the most general transformation $\mathbf{A} : \mathbf{x} \to \mathbf{x}'$ by writing the matrix as the triple product of separate transformations, each having a simple form. This is similar to what one does in finding the most general 3-dimensional rotation using the Euler angles. One then has

$$\mathbf{x}' = \mathbf{A}\mathbf{x}, \quad \mathbf{A} = \mathbf{B}\mathbf{C}\mathbf{D}, \qquad (3.17)$$

with $\mathbf{x}'^t \eta \mathbf{x}' = \mathbf{x}^t \eta \mathbf{x}$ and η is the Minkowskian metric in three dimensions whose diagonal is $(+1, -1, -1)$. The three matrices are then given by

$$\mathbf{B} = \begin{pmatrix} \gamma_1 & -\beta_1\gamma_1 & 0 \\ -\beta_1\gamma_1 & \gamma_1 & 0 \\ 0 & 0 & 1 \end{pmatrix}; \qquad (3.18a)$$

$$\gamma_1 = \left(1 - \beta_1^2\right)^{-1/2}, \quad \beta_1 = V_1/c, \qquad (3.18b)$$

$$\mathbf{C} = \begin{pmatrix} \gamma & 0 & -\gamma\beta \\ 0 & 1 & 0 \\ -\gamma\beta & 0 & \gamma \end{pmatrix}; \qquad (3.19a)$$

$$\gamma = \left(1 - \beta^2\right)^{-1/2}, \ \beta = T/\tau, \qquad (3.19b)$$

$$\mathbf{D} = \begin{pmatrix} \gamma_2 & -\beta_2\gamma_2 & 0 \\ -\beta_2\gamma_2 & \gamma_2 & 0 \\ 0 & 0 & 1 \end{pmatrix}; \qquad (3.20a)$$

$$\gamma_2 = \left(1 - \beta_2^2\right)^{-1/2}, \ \beta_2 = V_2/c. \qquad (3.20b)$$

The product matrix **A=BCD** then follows as

$$\mathbf{A} = \begin{pmatrix} \gamma_1\gamma_2\left(\gamma + \beta_1\beta_2\right) & -\gamma_1\gamma_2\left(\beta_2\gamma + \beta_1\right) & -\beta\gamma_1\gamma \\ -\gamma_1\gamma_2\left(\beta_1\gamma + \beta_2\right) & \gamma_1\gamma_2\left(\beta_1\beta_2\gamma + 1\right) & \beta_1\beta\gamma_1\gamma \\ -\beta\gamma\gamma_2 & \beta\beta_2\gamma\gamma_2 & \gamma \end{pmatrix}. \qquad (3.21)$$

3.5 Concluding remarks

It will be a great challenge to find out the covariant field equations under the group O(3,4), the extension of the Lorentz group to cosmology, presented in this chapter. Such a task was carried out by E.P. Wigner [3] and V. Bargmann [4] for the inhomogeneous Lorentz group who obtained the linear field equations associated to the representations of that group. These equations follow to be exactly the Maxwell, Dirac, Klein-Gordon, Proca, and Weyl equations. For our group, one should obtain extended versions of these equations to cosmology with terms of the order $1/\tau = H_0$ which are locally negligible but are not so globally. It is highly probable that these terms will accurately describe dark matter in a natural way.

Finally, the mathematical study, in particular the infinite-dimensional representations [5] of the groups presented here is a big task for mathematicians.

3.6 References

[1] M. Carmeli, *Found. Phys.* **25**, 1029 (1995); **26**, 413 (1996).
[2] M. Carmeli, *Commun. Theor. Phys.* **4**, 109 (1995).
[3] E.P. Wigner, *Ann. Math.* **40**, 149 (1939).
[4] V. Bargmann, *Ann. Math.* **48**, 568 (1947).
[5] M. Carmeli, *Group Theory and General Relativity* (McGraw-Hill, New York, 1977).

Chapter 4

Fundamentals of Einstein's Special Relativity

In this chapter and the rest of the book the fundamentals of Einstein's special relativity theory are presented. These fundamentals underline all the physical laws of nature which do not involve the gravitational field. The present chapter starts with the postulates of special relativity, namely the principles of relativity and the constancy of the speed of light. This is then followed by discussing the basic concept of coordinate systems, and particularly the inertial system. Simultaneity, an essential notion in special relativity theory, is subsequently analysed. These basic concepts are then followed by the Galilean transformation and group, and the Lorentz transformation and group. Consequences of the Lorentz transformation are then drawn. A four-dimensional formulation of spacetime, following Minkowski, is subsequently given in Chapter 5. The light cone structure, an important description of spacetime, is then given in Chapter 6. The discussion on special relativity theory is concluded in the last chapter by giving the relationship among the mass, energy and momentum along with the introduction of the energy-momentum four-vector. Our presentation will be along the original lines of Einstein's theory rather than dealing with technicalities, in order to emphasize the deep analogy of special relativity to the cosmological special relativity given in Chapter 2.

4.1 Postulates of special relativity

In the following we give the basic principles of the special theory of relativity. These principles are needed to describe the electromagnetic field and other physical phenomena, and they constitute their spacetime symmetry background [1-8].

The special theory of relativity was developed by Einstein in 1905 in order to overcome and correct certain basic concepts that were in use at that time, such as asymmetries in relative motion of bodies. Examples of relative motion in electrodynamics, and the unsuccessful attempt to detect the motion of the Earth by the experiment of Michelson and Morley, suggested that the phenomena of electrodynamics and mechanics do not depend on the Newtonian notion of absolute rest. Rather, the laws of electrodynamics should be valid in all frames of references in which the equations of mechanics are valid.

4.1.1 The principle of relativity. Constancy of the speed of light

Einstein raised the above observation to the status of a postulate and called it the *principle of relativity*. He also introduced

Postulates of special relativity

another postulate (which is only apparently inconsistent with the former one) according to which light always propagates in empty space with a constant velocity c which is independent of the motion of the emitting body and the measuring instrument.

The above two postulates were shown by Einstein to be enough for the development of a consistent theory of electrodynamics of moving charges which is based on Maxwell's original theory that was assumed to be valid in stationary systems only. The theory did not require an "absolute stationary space."

4.1.2 Coordinates

To describe the electromagnetic field, or any other classical field, one needs a system of coordinates in terms of which the fields are described. Such a coordinate system will include three *spatial* coordinates to which we add the *time* coordinate. The three spatial coordinates will be denoted by x^k, where lower case Latin indices $k = 1, 2, 3$, and the time coordinate by $x^0 = ct$, where c is the speed of light in vacuum. The four coordinates will collectively be denoted by x^α, where Greek indices take the values $\alpha = 0, 1, 2, 3$.

4.1.3 Inertial coordinate system

A system of coordinates in which the law of inertia holds is called an *inertial coordinate system*. Hence Newton's laws of mechanics are valid only in inertial coordinate systems.

If K is an inertial coordinate system, then every other coordinate system K' is also an inertial system if it is in uniform motion with respect to K. Hence if, relative to K, K' is a uniformly moving coordinate system then the physical laws can be expressed with respect to K' exactly as with respect to K.

It will be seen in the sequel that one of the most important physical consequences of the special relativity theory is the existence of a maximum signal speed in nature, which coincides with the velocity of light in empty space. It is therefore natural to define the same time at separate points by means of light signals. This then raises the problem of defining simultaneity.

4.1.4 Simultaneity

The definition of simultaneity is made as follows. If light requires the same time to pass across a path $A \to M$ as for a path $B \to M$, where M is in the middle of the distance AB, then we say that the light signals at A and B started simultaneously if the observer at M sees the two light signals at the same time.

Will two events, which occur simultaneously in one system, also be simultaneous in another system moving with a velocity v with respect to the first one? The answer is *negative*; events which are simultaneous in one coordinate system are not necessarily simultaneous in others. This is so since every inertial system has its own particular time.

4.2 The Galilean transformation

We have seen that inertial coordinate systems are those which are in uniform, rectilinear, translational motions with respect to each other. Accordingly, inertial systems of coordinates differ from each other by orthogonal rotations, accompanied by translations of the origins of the systems, and by motion with uniform velocities. One can, furthermore, add the translation of the time coordinate thus enabling an arbitrary choice of the origin of time $t = 0$.

Counting the number of parameters which each system of coordinates has with respect to any other, we find that there are ten.

4.2.1 The Galilean group

A transformation between inertial coordinate systems which has ten parameters, as described above, is called a *Galilean transformation*. The aggregate of all Galilean transformations provides a group, called the *Galilean group*, which has ten parameters.

One can choose two inertial systems of coordinates so that their corresponding axes are parallel and coincide at $t = 0$. If v is the velocity of one inertial coordinate system with respect to the other, the Galilean transformation can then be reduced to a simple transformation as follows:

$$x' = x - v_x t, \quad y' = y \quad v_y t, \quad z' = z - v_z t, \qquad (4.1)$$

where v_x, v_y and v_z are the components of the velocity **v** along the x axis, y axis, and z axis, respectively. Of course the Newtonian laws of classical mechanics are invariant under the full ten-parameter Galilean group of transformations, and we have what can be called a *Galilean invariance*.

Finally we mention that if we have three systems of coordinates which move with constant velocities with respect to each other, then one can obtain the velocity between any two systems by adding or subtracting the appropriate relative velocities of the three systems. We shall see later on that this result, which expresses the law of the addition of velocities in classical mechanics, is valid only for velocities much smaller than that of light in the special theory of relativity.

In the next section the Lorentz transformation, a generalization of the Galilean transformation, will be derived.

4.3 The Lorentz transformation

In the last section we presented the (nonrelativistic) Galilean transformation of spatial coordinates. We are now in a position

42   Chapter 4. Fundamentals of Einstein's Special Relativity

to generalize this transformation to the relativistic case, where both the time and the spatial coordinates are involved.

4.3.1   Measuring rods and clocks

We notice that the assumptions of the existence of measuring rods and clocks are *not* independent of each other. This is so since a light signal, which is reflected back and forth, may provide an ideal clock, remembering that the speed of light in vacuum is constant. (For more details see Bondi [6].)

4.3.2   Spatial coordinates and time

We assume that there exists a rigid body of reference which is moving, and thus provides an inertial coordinate system. In this system, the spatial coordinates then denote the results of measurements that are made with stationary rods.

The time of an event can also be made in an analogous way. One then needs a way to measure the time differences by a periodic process. A clock at rest in an inertial coordinate system records a local time. The local times of all points of an inertial coordinate system then give *the* time of that system, assuming that the clocks are at rest relative to each other. The times of different inertial coordinate systems are not necessarily identical if light is used to synchronize the clocks.

4.3.3   Einstein's paradox

The prerelativity difficulty with light was beautifully illustrated by Einstein ("a paradox upon which I had already hit at the age of sixteen") by the following *gedanken* experiment: If an observer follows a light beam with the velocity c, the beam would be observed as an electromagnetic field at rest which is spatially

The Lorentz transformation

oscillating. However, this is impossible by Maxwell's equations. In fact, such an observer would see the same as another one who is not moving at all. The above difficulty exemplifies the essence of the special relativity theory. The difficulty is caused by the prerelativistic assumption of the absolute time.

4.3.4 Apparent incompatibility of the special relativity postulates

The above difficulty can be formulated in a different way as follows.

According to the Galilean transformation which relates the spatial coordinates and the time between inertial systems in prerelativity, the postulates of the constancy of the speed of light and of the principle of relativity (which applies, in particular, to the propagation of light and hence its constant velocity is independent of the choice of inertial system) are mutually incompatible, even though both are experimentally valid.

The special theory of relativity resolves this impasse as follows.

The above two postulates will be compatible with each other if a new transformation relating the spatial coordinates and times of different inertial systems replaces the Galilean transformation. The new transformation, of course, follows to be the Lorentz transformation. This, subsequently, requires certain behavior of the moving measuring rods and clocks.

The principle of relativity may, thus, alternatively be restated as follows: *The laws of physics should be covariant (or invariant) under the Lorentz transformations relating different inertial coordinate systems.* This Lorentz invariance is in accordance with the Michelson-Morley null experiment which showed that on the moving Earth light spreads with the same speed in all directions.

Consequently, the behavior of light is not incompatible with the principle of relativity. The incompatibility is only apparent.

Before we derive the Lorentz transformation, the following remark is worth mentioning.

4.3.5 Remark on action-at-a-distance

In spite of the fact that the Newtonian notion of action-at-a-distance is not allowed by special relativity, the theory does permit the introduction of an action-at-a-distance which propagates with the speed of light. And indeed such a task was carried out by Van Dam and Wigner [9]. Such an action-at-a-distance, however, would mean the abandon of the concept of fields which is the basis of modern physics, and thus it is of little interest.

4.3.6 Derivation of the Lorentz transformation

We now return to the problem of apparent disagreement between the law of propagation of light in vacuum and the principle of relativity, and how to resolve it. This then leads to the following question: Given the spatial coordinates and time of an event in an inertial coordinate system K, what are the corresponding quantities of the same event in another inertial system K', taking into account that light rays propagate with the same speed c in both systems? The answer to this question leads to the Lorentz transformation determining the values t', x', y', z' of an event with respect to K' from the corresponding magnitudes t, x, y, z of the same event with respect to K, when the law of propagation of light is satisfied in both systems K and K'.

Consider two inertial coordinate systems K and K' whose origins coincide at time $t = 0$, and the events with respect to which are denoted by t, x, y, z and t', x', y', z', respectively. A light pulse emitted from the origin of K will be spread spherically

The Lorentz transformation

with the speed c, according to the equation

$$x^2 + y^2 + z^2 = c^2 t^2. \tag{4.2}$$

Invariance of the speed of light tells us that an observer in K' will also see the light propagating from his origin spherically according to the equation

$$x'^2 + y'^2 + z'^2 = c^2 t'^2. \tag{4.3}$$

From Eqs. (4.2) and (4.3) one then obtains

$$c^2 t'^2 - \left(x'^2 + y'^2 + z'^2\right) = c^2 t^2 - \left(x^2 + y^2 + z^2\right), \tag{4.4}$$

or

$$\eta_{\mu\nu} x'^\mu x'^\nu = \eta_{\mu\nu} x^\mu x^\nu, \tag{4.5}$$

where x^μ and x'^μ are defined by

$$x^\mu = (ct,\ x,\ y,\ z),\quad x'^\mu = (ct',\ x',\ y',\ z'), \tag{4.6}$$

and the symbol $\eta_{\mu\nu}$ (and later on $\eta^{\mu\nu}$) is the flat-space metric, given by the matrix

$$\eta = \begin{pmatrix} +1 & 0 & 0 & 0 \\ 0 & -1 & 0 & 0 \\ 0 & 0 & -1 & 0 \\ 0 & 0 & 0 & -1 \end{pmatrix}. \tag{4.7}$$

In the above equations, and throughout the following, repeated indices indicate the use of the summation convention.

We will seek a linear transformation of the form

$$x'^\mu = \Lambda^\mu{}_\nu x^\nu \tag{4.8}$$

between the times and spatial coordinates of the two inertial systems K and K'. Using matrix notation, Eqs. (4.5) and (4.8) can then be written in the form

$$x'^t \eta x' = x^t \eta x \tag{4.9}$$

and
$$x' = \Lambda x, \tag{4.10}$$
respectively. Here x and x' are the one-column matrices
$$x = \begin{pmatrix} x^0 \\ x^1 \\ x^2 \\ x^3 \end{pmatrix}, \quad x' = \begin{pmatrix} x'^0 \\ x'^1 \\ x'^2 \\ x'^3 \end{pmatrix}, \tag{4.11}$$
x^t and x'^t are the transposed matrices to the matrices x and x', respectively, and Λ is the 4×4 matrix whose elements are $\Lambda^\mu{}_\nu$.

Using now Eq. (4.10) in Eq. (4.9) then gives
$$x^t \Lambda^t \eta \Lambda x = x^t \eta x, \tag{4.12}$$
from which we obtain the condition
$$\Lambda^t \eta \Lambda = \eta \tag{4.13}$$
that the 4×4 matrix Λ of the Lorentz transformation has to satisfy. Equation (4.13) is a generalization of the familiar relation
$$R^t I R = I, \tag{4.14}$$
which the 3×3 orthogonal matrix R, describing ordinary rotations of the spatial coordinates alone, satisfies. The essential difference between the two cases is in the replacement of the unit matrix I in the ordinary three-dimensional rotations by the matrix η in the four-dimensional Lorentz transformations.

Hence the transformation we are looking after is a "rotation" in a four-dimensional spacetime which consists of the time and the three dimensions of the ordinary space. Such a spacetime is usually called the *Minkowskian spacetime.*

The Lorentz transformation is thus the "orthogonal" transformation of the Minkowskian spacetime.

The Lorentz transformation

We now derive the Lorentz transformation connecting the two coordinate systems K and K' when they have the same orientations and their origins coincide at $t = 0$, but K' moves along the coordinate x with a velocity v as shown in Figure 4.1.

The directions perpendicular to the motion are obviously left unaffected by the transformation. Hence

$$x'^2 = x^2, \quad x'^3 = x^3, \tag{4.15}$$

and only the x^0 and x^1 coordinates require changes when transforming from one system to the other. One will therefore have the form

$$\Lambda = \begin{pmatrix} \Lambda^0{}_0 & \Lambda^0{}_1 & 0 & 0 \\ \Lambda^1{}_0 & \Lambda^1{}_1 & 0 & 0 \\ 0 & 0 & 1 & 0 \\ 0 & 0 & 0 & 1 \end{pmatrix} \tag{4.16}$$

for the matrix of the Lorentz transformation in our particular case.

The "orthogonality" condition (4.13) then yields the equation

$$\eta_{CD} \Lambda^C{}_A \Lambda^D{}_B = \eta_{AB}, \tag{4.17}$$

where the indices A, B, C, $D = 0, 1$, and $\eta_{00} = -\eta_{11} = 1$, $\eta_{01} = \eta_{10} = 0$. The above formula gives three relations connecting the four elements of the matrix (4.16):

$$\left(\Lambda^0{}_0\right)^2 - \left(\Lambda^1{}_0\right)^2 = 1,$$

$$\left(\Lambda^0{}_1\right)^2 - \left(\Lambda^1{}_1\right)^2 = -1, \tag{4.18}$$

$$\Lambda^0{}_0 \Lambda^0{}_1 - \Lambda^1{}_0 \Lambda^1{}_1 = 0.$$

The solution of these equations can therefore be determined up to an arbitrary parameter. One then finds that

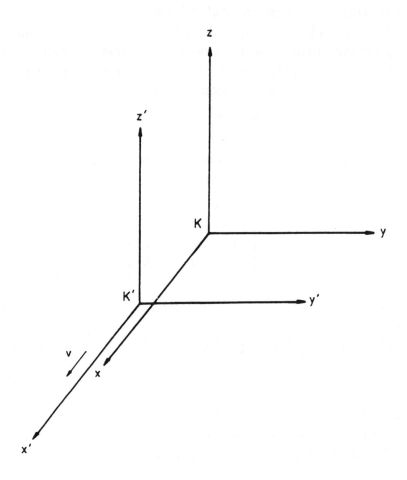

Figure 4.1 Two coordinate systems K and K', one moving with respect to the other with a velocity v in the x-direction.

The Lorentz transformation

$$\Lambda^0{}_0 = \cosh\psi, \quad \Lambda^0{}_1 = \sinh\psi,$$
$$\Lambda^1{}_0 = \sinh\psi, \quad \Lambda^1{}_1 = \cosh\psi, \qquad (4.19)$$

is such an appropriate solution. With these values for the four elements, we obtain

$$\Lambda = \begin{pmatrix} \cosh\psi & \sinh\psi & 0 & 0 \\ \sinh\psi & \cosh\psi & 0 & 0 \\ 0 & 0 & 1 & 0 \\ 0 & 0 & 0 & 1 \end{pmatrix} \qquad (4.20)$$

for the matrix (4.16) of the Lorentz transformation.

The parameter ψ is related to the relative velocity v between the two inertial coordinate systems K and K'. The relationship between them is found by determining the motion of the origin of the coordinate system K as seen from K', for instance. This motion is determined by putting $x^1 = 0$ in the Lorentz transformation (4.10), using Eqs. (4.11) and (4.20). This gives

$$x'^0 = x^0 \cosh\psi,$$
$$x'^1 = x^0 \sinh\psi. \qquad (4.21)$$

We therefore obtain

$$\frac{x'^1}{x'^0} = \frac{1}{c}\frac{x'}{t'} = -\beta = \tanh\psi, \qquad (4.22)$$

where the parameter β is defined by

$$\beta = \frac{v}{c}. \qquad (4.23)$$

Accordingly we obtain from Eq. (4.22)

$$\cosh\psi = \frac{1}{\sqrt{1-\beta^2}}, \qquad (4.24a)$$

$$\sinh\psi = \frac{-\beta}{\sqrt{1-\beta^2}}. \qquad (4.24b)$$

Chapter 4. Fundamentals of Einstein's Special Relativity

Using these results in Eq. (4.20) then yields

$$\Lambda = \begin{pmatrix} \dfrac{1}{\sqrt{1-\beta^2}} & \dfrac{-\beta}{\sqrt{1-\beta^2}} & 0 & 0 \\ \dfrac{-\beta}{\sqrt{1-\beta^2}} & \dfrac{1}{\sqrt{1-\beta^2}} & 0 & 0 \\ 0 & 0 & 1 & 0 \\ 0 & 0 & 0 & 1 \end{pmatrix} \qquad (4.25)$$

for the matrix of the Lorentz transformation. We also obtain

$$\Lambda^{-1} = \begin{pmatrix} \dfrac{1}{\sqrt{1-\beta^2}} & \dfrac{\beta}{\sqrt{1-\beta^2}} & 0 & 0 \\ \dfrac{\beta}{\sqrt{1-\beta^2}} & \dfrac{1}{\sqrt{1-\beta^2}} & 0 & 0 \\ 0 & 0 & 1 & 0 \\ 0 & 0 & 0 & 1 \end{pmatrix} \qquad (4.26)$$

for the inverse matrix describing the inverse Lorentz transformation.

The Lorentz transformation along the x axis is therefore given by

$$ct' = \frac{ct - \beta x}{\sqrt{1-\beta^2}}, \qquad (4.27a)$$

$$x' = \frac{x - \beta ct}{\sqrt{1-\beta^2}}, \qquad (4.27b)$$

$$y' = y, \ z' = z. \qquad (4.27c)$$

We also obtain

$$ct = \frac{ct' + \beta x'}{\sqrt{1-\beta^2}}, \qquad (4.28a)$$

$$x = \frac{x' + \beta ct'}{\sqrt{1-\beta^2}}, \qquad (4.28b)$$

$$y = y', \ z = z', \qquad (4.28c)$$

The Lorentz transformation

for the inverse transformation from the coordinates x'^μ back to x^μ. Different proofs to Eqs. (4.27) are given in Problems 4.3.1 and 4.3.2. They can also be proved by using an analogous method to that used in Section 2.11 for deriving the cosmological transformation.

Equations (4.28) show that the inverse transformation differs from Eqs. (4.27) only by a change in the sign of v. This result is obvious since the coordinate system K is moving relative to the system K' with the velocity $-v$.

A Lorentz transformation involving the time coordinate x^0 and one or more spatial coordinates x^k, such as that derived above, is often called a *boost*. A Lorentz transformation which keeps the time coordinate unchanged is, of course, just an ordinary three-dimensional rotation of the spatial coordinates.

In the same way one can find the other Lorentz transformations along the y axis and the z axis.

Finally we notice that if one neglects nonlinear terms in v/c in the Lorentz transformation (4.27), one obtains the *approximate Lorentz transformation* (see Problem 4.3.3):

$$\begin{aligned} x'^0 &= x^0 - \beta x^1, \\ x'^1 &= x^1 - \beta x^0, \\ x'^2 &= x^2, \\ x'^3 &= x^3, \end{aligned} \qquad (4.29)$$

in which the coordinates x^0 and x^1 appear on the same footing.

If one, in addition, neglects the term with β in the first of the above equations, one then obtains the nonrelativistic transformation

$$x' = x - vt, \quad y' = y, \quad z' = z, \quad t' = t. \qquad (4.30)$$

Here the time and the coordinate x do not appear on the same footing.

Equations (4.30), of course, describe a Galilean transformation along the x axis. It is a particular case of the Galilean transformation, discussed in Section 4.2 and given by Eqs. (4.1), for which $v_x = v$ and $v_y = v_z = 0$.

4.3.7 The Lorentz group

To conclude this section we give a brief discussion on the groups which can be obtained from the Lorentz transformations. (For representations of the Lorentz group see [10].)

The Lorentz transformations form a group called the (homogeneous) Lorentz group. It is a subgroup of the *inhomogeneous Lorentz group*, also known as the *Poincaré group*. The latter group is formed from the *inhomogeneous* Lorentz transformations

$$x'^\mu = \Lambda^\mu{}_\nu x^\nu + x_0^\mu, \qquad (4.31)$$

where x_0^μ describes *translations*.

The Lorentz group possesses four disconnected parts which arise as follows.

Equation (4.13) shows that $(\det\Lambda)^2 = 1$. Accordingly, the determinant of every Lorentz transformation is equal to either $+1$,

$$\det\Lambda = +1, \qquad (4.32)$$

in which case the transformation is called *proper*, or to -1,

$$\det\Lambda = -1, \qquad (4.33)$$

in which it is called *improper*.

From Eq. (4.13) when written with indices,

$$\eta_{\mu\nu} \Lambda^\mu{}_\alpha \Lambda^\nu{}_\beta = \eta_{\alpha\beta}, \qquad (4.34)$$

and taking $\alpha = \beta = 0$, one obtains

$$\left(\Lambda^0{}_0\right)^2 - \left(\Lambda^1{}_0\right)^2 - \left(\Lambda^2{}_0\right)^2 - \left(\Lambda^3{}_0\right)^2 = 1. \qquad (4.35)$$

The Lorentz transformation

Therefore $(\Lambda^0{}_0)^2 \geq 1$, and consequently we have either

$$\Lambda^0{}_0 \geq +1, \tag{4.36}$$

in which case the transformation is called *orthochronous*, or

$$\Lambda^0{}_0 \leq -1. \tag{4.37}$$

The aggregate of all orthochronous Lorentz transformations provides a subgroup of the Lorentz group.

The four parts of the Lorentz group are described as follows:
(1) $L_+^\uparrow$: $\det\Lambda = +1$, $\Lambda^0{}_0 \geq +1$. This part contains the identity element of the group. The aggregate of all proper, orthochronous, Lorentz transformations provides a group, which is a subgroup of the Lorentz group. It is called the *proper, orthochronous, Lorentz group*.
(2) $L_-^\uparrow$: $\det\Lambda = -1$, $\Lambda^0{}_0 \geq +1$. This part contains a *space inversion* element S which describes a reflection relative to the three spatial axes:

$$x'^0 = x^0, \quad x'^1 = -x^1, \quad x'^2 = -x^2, \quad x'^3 = -x^3. \tag{4.38}$$

(3) $L_-^\downarrow$: $\det\Lambda = -1$, $\Lambda^0{}_0 \leq -1$. This part contains a *time reversal* element T which describes a reflection relative to the time axis:

$$x'^0 = -x^0, \quad x'^1 = x^1, \quad x'^2 = x^2, \quad x'^3 = x^3. \tag{4.39}$$

(4) $L_+^\downarrow$: $\det\Lambda = +1$, $\Lambda^0{}_0 \leq -1$. This part contains the element ST.

As was mentioned before, from the above four parts of the Lorentz group one obtains the subgroup $L^\uparrow = L_+^\uparrow \cup L_-^\uparrow$ (the union of $L_+^\uparrow$ and $L_-^\uparrow$), called the *orthochronous Lorentz group*. Likewise, the subgroup $L_+ = L_+^\uparrow \cup L_+^\downarrow$, called the *proper Lorentz group*, is obtained.

54   Chapter 4. Fundamentals of Einstein's Special Relativity

Finally, we notice that every improper Lorentz transformation can be written in the form

$$\Lambda = S\Lambda_p, \qquad (4.40)$$

where S is a space-inversion element and Λ_p is a proper Lorentz transformation.

In the next section some important consequences of the Lorentz transformation are drawn.

4.3.8   Problems

4.3.1 Let a light ray be emitted from the origin of a moving system K' (see Figure 4.1) at the time t'_0 along the x axis to $\tilde{x} = x - vt$, where v is the velocity of K' relative to the "stationary" system K. Let the ray then be reflected back at time t'_1 to the origin of the coordinates, arriving there at time t'_2. We then have

$$\frac{1}{2}(t'_0 + t'_2) = t'_1. \qquad (1)$$

Derive the Lorentz transformation (4.27) by inserting in Eq. (1) the arguments of the function t', and applying the postulate of the constancy of the speed of light in the system K. (This is Einstein's historical derivation of the Lorentz transformation. The solution given below is a concise version of the original one [1].)

Solution: The function t' depends on the four coordinates of the system K, namely $t' = t'(t, x, y, z)$. Inserting these arguments in Eq. (1), and using the postulate of the constancy of the speed of light in the system K, one obtains after a lengthy but straightforward calculation,

$$\frac{v}{c^2 - v^2}\frac{\partial t'}{\partial t} + \frac{\partial t'}{\partial \tilde{x}} = 0.$$

The Lorentz transformation

Likewise, one obtains

$$\frac{\partial t'}{\partial y} = 0, \quad \frac{\partial t'}{\partial z} = 0, \tag{3}$$

since light always propagates along the y and z axes, when viewed from the stationary system K, with speed $(c^2 - v^2)^{1/2}$.

If one now assumes, furthermore, that t' is a *linear* function of its arguments, and using Eq. (2), then one finds

$$t' = a\left(t - \frac{v}{c^2 - v^2}\tilde{x}\right), \tag{4}$$

where a is a function of v (at present unknown), and it is assumed that at the origin of K', $t' = 0$ when $t = 0$.

Using Eq. (4) one can then determine the coordinates x', y', z', taking into account the fact that light propagates with the speed c when measured in the moving system K'. For a ray of light emitted at time $t' = 0$ along the x' axis, one accordingly has

$$x' = ct' = ca\left(t - \frac{v}{c^2 - v^2}\tilde{x}\right). \tag{5}$$

But the ray satisfies the propagation equation $x - ct$, and therefore

$$\tilde{x} = x - vt = (c - v)t, \tag{6}$$

or

$$t = \frac{\tilde{x}}{c - v}. \tag{7}$$

Using now this expression for t in Eq. (5) one obtains

$$x' = \frac{c^2 a \tilde{x}}{c^2 - v^2}. \tag{8}$$

Likewise, one can determine the coordinate y' and z' by considering rays moving along them. One obtains

$$y' = \frac{cay}{\sqrt{c^2 - v^2}},$$

Chapter 4. Fundamentals of Einstein's Special Relativity

$$z' = \frac{caz}{\sqrt{c^2 - v^2}}. \tag{9b}$$

Equations (4), (8) and (9) express the dependence of the coordinates t', x', y', z' on t, x, y, z, provided one substitutes for $\tilde{x}$ its value $x - vt$.

A straightforward calculation then gives:

$$\begin{aligned} t' &= \phi(v)\gamma(v)\left(t - \frac{vx}{c^2}\right), \\ x' &= \phi(v)\gamma(v)(x - vt), \\ y' &= \phi(v)y, \\ z' &= \phi(v)z. \end{aligned} \tag{10}$$

Here $\gamma(v)$ is defined by

$$\gamma(v) = \frac{1}{\sqrt{1 - \frac{v^2}{c^2}}}, \tag{11}$$

and $\phi(v)$ is a new unknown function of v, related to the function a (of v also) by

$$\phi(v) = \frac{a}{\sqrt{1 - \frac{v^2}{c^2}}}. \tag{12}$$

The function $\phi(v)$ is left, as yet, arbitrary.

In order to determine $\phi(v)$ one introduces a third coordinate system K'' which is moving with the velocity $-v$ relative to K', and writes the transformation law of its coordinates t'', x'', y'', z'' in terms of those of K and K'. One then obtains, after a lengthy but straightforward calculation,

$$\phi(v) = 1, \tag{13}$$

The Lorentz transformation

and
$$t' = \gamma\left(t - \frac{vx}{c^2}\right),$$
$$x' = \gamma(x - vt), \tag{14}$$
$$y' = y,$$
$$z' = z,$$

for the transformation of the time and the spatial coordinates. In the above formulas γ is given by Eq. (11).

Equations (14) are, of course, those of the Lorentz transformation, Eqs. (4.27), given in the text.

4.3.2 Derive the Lorentz transformation by considering the transmission of light signals along the positive and negative parts of the x axis. (Also given by Einstein [2]. The solution given below is a concise version of the original one.)

Solution: One expresses the coordinates x' and t' in terms of x and t, where unprimed and primed quantities refer to the systems K and K', respectively, of Figure 4.1.

Let a light signal be transmitted along the positive x axis. Then its equation of propagation is given by

$$x = ct, \tag{1}$$

where c is the speed of light in vacuum. Relative to the system K', the light signal also propagates with the speed c. Accordingly,

$$x' - ct' = 0 \tag{2}$$

represents the propagation of light in the system K'. Spacetime events which satisfy Eq. (1) must also satisfy Eq. (2). This will, indeed, be the case if a relation of the form

$$(x' - ct') = \lambda(x - ct) \tag{3}$$

is fulfilled, where λ is a constant. Equation (3) shows that the vanishing of $(x - ct)$ yields the vanishing of $(x' - ct')$.

Chapter 4. Fundamentals of Einstein's Special Relativity

Light rays transmitted along the negative x axis, likewise, satisfy

$$(x' + ct') = \mu (x + ct), \qquad (4)$$

where μ is a constant. From Eqs. (3) and (4) one then obtains a linear transformation between the variables ct and x,

$$ct' = act - bx, \qquad (5a)$$

$$x' = ax - bct, \qquad (5b)$$

where a and b are two new constants related to λ and μ by

$$a = \frac{1}{2} (\lambda + \mu), \qquad (6a)$$

$$b = \frac{1}{2} (\lambda - \mu). \qquad (6b)$$

It remains to determine the constants a and b in terms of the relative velocity v between the two systems K and K'.

A lengthy, but straightforward, calculation then yields

$$a = \frac{1}{\sqrt{1 - \frac{v^2}{c^2}}}, \quad b = \frac{v/c}{\sqrt{1 - \frac{v^2}{c^2}}}. \qquad (7)$$

Using these expressions for the constants a and b in Eqs. (5) then gives

$$ct' = \frac{ct - \beta x}{\sqrt{1 - \beta^2}}, \quad x' = \frac{x - \beta ct}{\sqrt{1 - \beta^2}}, \qquad (8a)$$

where $\beta = v/c$. When supplemented by the relations

$$y' = y, \quad z' = z, \qquad (8b)$$

one thus obtains the Lorentz transformation which was given in the text by Eqs. (4.27).

4.3.3 Find a nonrelativistic approximation to the Lorentz transformation which, unlike the Galilean one, contains at the same footing both the spatial coordinates and time.

Solution: Such an approximate transformation, when neglecting nonlinear terms in v/c, is given by

$$\begin{aligned} x'^0 &= x^0 - \frac{v}{c}x^1, \\ x'^1 &= x^1 - \frac{v}{c}x^0, \\ x'^2 &= x^2, \quad x'^3 &= x^3, \end{aligned} \quad (1)$$

if the coordinates x^2 and x^3 are kept untransformed. Here $x^0 = ct$, $x^1 = x$, $x^2 = y$, $x^3 = z$, and c is the speed of light in vacuum.

One then finds that (neglecting nonlinear terms in v/c)

$$c^2 t'^2 - \left(x'^2 + y'^2 + z'^2\right) = c^2 t^2 - \left(x^2 + y^2 + z^2\right). \quad (2)$$

Moreover, the aggregate of all the transformations (1) provides a group (keeping only linear terms in v/c).

4.4 Consequences of the Lorentz transformation

We now draw some important consequences from the Lorentz transformation derived in the last section. These will include the Lorentz contraction of lengths, the dilation of time scales, and the law of addition of velocities. More implications of the Lorentz transformation will be discussed in the sequel.

4.4.1 Nonrelativistic limit

In the limit of small velocities relative to the speed of light c, namely $\beta \ll 1$, Eqs. (4.27) are easily seen to be reduced to the

Galilean transformation.

4.4.2 The Lorentz contraction of lengths

Consider a rod whose length is one unit, and placed along the x' axis of the system K' between the points $x' = 0$ and $x' = 1$. What is the length of this rod relative to the system K?

The above question is answered by determining the ends of the rod in K at a particular time t of K. Taking $t = 0$, for instance, and using the second formula of Eqs. (4.27), we obtain

$$x = x'\sqrt{1-\beta^2}; \quad \beta = v/c. \qquad (4.41)$$

The values of the ends of the rod in K are then obtained from Eq. (4.41) by putting $x' = 0$ and $x' = 1$. This gives $x = 0$ and $x = \sqrt{1-\beta^2}$, respectively. The length of the rod in K is thus $\sqrt{1-\beta^2}$, rather than unit.

But the rod is moving relative to K with the velocity v. Hence the length of a rod moving with the velocity v is seen to be contracted to $\sqrt{1-\beta^2}$ times its length; it is shorter when in motion than when at rest, a result known as the *Lorentz contraction of length*.

On the contrary, if the rod would have been placed at rest along the x axis in the coordinate system K, its length would also be contracted by the same factor $\sqrt{1-\beta^2}$ as seen from K'. This result is, of course, in accordance with the principle of relativity.

Of course no contraction effect can be obtained using the Galilean transformation.

It will be recalled that the phenomenon of the length contraction exists also in the cosmological special relativity as discussed in Subsection 2.15.2.

4.4.3 The dilation of time

The dilation of time scale can, similarly, be shown as follows.

Let a clock be placed at the origin ($x' = 0$) of the coordinate system K', and let $t' = 0$ and $t' = 1$ be two of its successive ticks. From the first two equations of the Lorentz transformation (4.27), when $x' = 0$, we obtain

$$t' = t\sqrt{1-\beta^2}; \quad \beta = v/c. \tag{4.42}$$

Hence for $t' = 0$ and $t' = 1$ we obtain

$$t = 0, \tag{4.43}$$

and

$$t = \frac{1}{\sqrt{1-\beta^2}}, \tag{4.44}$$

respectively.

As observed from the coordinate system K, the clock is moving with the velocity v, and the time which elapses between two of its successive strokes is not one second but $1/\sqrt{1-\beta^2}$ seconds, namely a longer time. Thus the clock goes *more slowly* when it is in motion than when it is at rest. Such a phenomenon is called the *dilation of time*.

The same conclusion would have been reached, of course, if the clock was placed in the system K and its time was judged from K'. Again the clock will be seen to run slower.

The analogous to this phenomenon in cosmological relativity is, of course, the velocity contraction given in Subsection 2.15.3.

4.4.4 The addition of velocities law

It might be thought possible to obtain a velocity greater than c by letting a particle move with a velocity w along the x' axis in K', and consider the motion from the system K with respect to which K' moves with the speed v along the x axis. Let the velocity of the particle be V with respect to K, and hence

$$x = Vt. \tag{4.45}$$

Classical mechanics, of course, gives $V = v + w$. However, this is not the case in special relativity. To see this we apply the Lorentz transformation, given by Eqs. (4.27), to the relation

$$x' = wt'. \tag{4.46}$$

A straightforward calculation then gives

$$x = \frac{(v+w)}{1 + \frac{vw}{c^2}} t = Vt, \tag{4.47}$$

and accordingly

$$V = \frac{v+w}{1 + \frac{vw}{c^2}}. \tag{4.48}$$

Equation (4.48) is called the *addition of velocities law*.

The same result is obtained by considering three coordinate systems, the second moves with the velocity v with respect to the first, and the third moves with the velocity w with respect to the second system. One may find the Lorentz transformation from the first to the third system directly by multiplying the matrices of the two separate transformations. It is then found that the total transformation corresponds to a velocity V given by Eq. (4.48), or equivalently by

$$\frac{V}{c} = \frac{\frac{v}{c} + \frac{w}{c}}{1 + \frac{vw}{c^2}}. \tag{4.49}$$

It is seen from Eq. (4.49) that V/c is always less than unity. The proof of Eq. (4.49) is left for the reader (see Problem 4.4.1).

Consequences of the Lorentz transformation

From the above discussion we also conclude that, in special relativity, the speed of light c is a limiting velocity. Namely, the velocity c can neither be reached nor exceeded by a finite-mass particle. The same conclusion clearly follows from the Lorentz transformation itself since it becomes meaningless for values of v larger than c.

The analogous to this law in cosmological special relativity is, of course, the law of addition of cosmic times given in Subsection 2.15.4.

In the next chapter the four-dimensional structure of space-time is discussed.

4.4.5 Problems

4.4.1 Prove the law of addition of velocities, Eq. (4.49), by considering two Lorentz transformations as successive rotations in the $x^0 - x^1$ plane.

Solution: The solution is left for the reader.

4.4.2 Derive the formulas relating the velocity of a particle in one inertial coordinate system K to that in a second system K', where K' moves relative to K with the velocity V along the x axis. Use Eqs. (4.27) to show that [11]

$$v'_x = \frac{v_x - V}{1 - \frac{V v_x}{c^2}}, \tag{1a}$$

$$v'_y = \frac{v_y \sqrt{1 - \frac{V^2}{c^2}}}{1 - \frac{V v_x}{c^2}}, \tag{1b}$$

$$v'_z = \frac{v_z \sqrt{1 - \frac{V^2}{c^2}}}{1 - \frac{V v_x}{c^2}}, \tag{1c}$$

where $v_x = dx/dt$, etc. and $v'_x = dx'/dt'$, etc.

Use Eqs. (4.28), likewise, to show that the inverse transformation is given by

$$v_x = \frac{v'_x + V}{1 + \dfrac{V v'_x}{c^2}}, \qquad (2a)$$

$$v_y = \frac{v'_y \sqrt{1 - \dfrac{V^2}{c^2}}}{1 + \dfrac{V v'_x}{c^2}}, \qquad (2b)$$

$$v_z = \frac{v'_z \sqrt{1 - \dfrac{V^2}{c^2}}}{1 + \dfrac{V v'_x}{c^2}}. \qquad (2c)$$

Solution: The solution is left for the reader.

4.4.3 Find the change in the direction of the velocity under the transition from one coordinate system K to another system K' moving with the velocity V with respect to K along the x axis. To this end, consider a particle moving in two dimensions. Then

$$v_x = v \cos\theta, \quad v_y = v \sin\theta, \qquad (1)$$

$$v'_x = v' \cos\theta', \quad v'_y = v' \sin\theta', \qquad (2)$$

for instance, in the two coordinate systems K and K'. Use Eqs. (1) and (2) in Eqs. (1) and (2) of the previous problem, and show that

$$\tan\theta' = \frac{v\sqrt{1-\beta^2}\sin\theta}{v\cos\theta - V}, \qquad (3a)$$

$$\tan\theta = \frac{v'\sqrt{1-\beta^2}\sin\theta'}{v'\cos\theta' + V}, \qquad (3b)$$

where $\beta = V/c$ [11].

Solution: The solution is left for the reader.

4.4.4
Apply the results of Problems 4.4.2 and 4.4.3 to derive the aberration of light formula [11].

Solution: The apparent change in the direction of propagation of light under the transition from one inertial coordinate system to another can be obtained, using Eqs. (1) and (2) of Problem 4.4.2 and Eqs. (1) and (2) of Problem 4.4.3, by taking $v = v' = c$. One then gets

$$\cos\theta' = \frac{\cos\theta - \beta}{1 - \beta\cos\theta}, \tag{1}$$

$$\cos\theta = \frac{\cos\theta' + \beta}{1 + \beta\cos\theta'}. \tag{2}$$

Here the angles θ and θ' refer to the coordinate systems K and K', respectively, and $\beta = V/c$, where V is the velocity of K' with respect to K.

Assuming now that the relative velocity between the two coordinate systems is much smaller than the speed of light, $\beta \ll 1$, one easily finds, using Eq. (1),

$$\cos\theta' \approx \cos\theta - \beta\sin^2\theta. \tag{3}$$

The *aberration angle* $\Delta\theta = \theta' - \theta$, using a straightforward calculation, is then given by

$$\Delta\theta \approx \cot\theta - \frac{\cos\theta'}{\sin\theta}. \tag{4}$$

Using Eq. (3) in Eq. (4) then gives

$$\Delta\theta \approx \beta\sin\theta = \frac{V}{c}\sin\theta, \tag{5}$$

which is the *aberration of light formula*.

4.5 References

[1] A. Einstein, *Ann. Physik* **17**, 891 (1905); English translation in: *The Principle of Relativity* (Dover, New York, 1923), p. 35.
[2] A. Einstein, *Relativity: The Special and General Theory* (Crown Publishers, New York, 1931).
[3] A. Einstein, *Autobiographical Notes*, P.A. Schilpp, Editor (Open Court Publishing Company, La Salle and Chicago, Illinois, 1979).
[4] D. Bohm, *The Special Theory of Relativity* (Benjamin, New York, 1965).
[5] M. Born, *Einstein's Theory of Relativity* (Dover, New York, 1962).
[6] H. Bondi, in: *1964 Brandeis Summer Institute in Theoretical Physics*, Vol. 1 (Prentice-Hall, Englewood Cliffs, New Jersey, 1965), pp. 379-406.
[7] A.I. Miller, *Albert Einstein's Special Theory of Relativity* (Addison-Wesley, Reading, Massachusetts, 1981).
[8] A.P. French, *Special Relativity* (W.W. Norton, New York and London, 1968).
[9] H. Van Dam and E.P. Wigner, *Phys. Rev.* **138**, B1576 (1965).
[10] M. Carmeli and S. Malin, *Representations of the Rotation and Lorentz Groups: An Introduction* (Marcel Dekker, New York and Basel, 1976).
[11] L.D. Landau and E.M. Lifshitz, *The Classical Theory of Fields* (Addison-Wesley, Reading, Massachusetts, 1959).

Chapter 5

Structure of Spacetime

In the last chapter the foundations of special relativity were presented. These were based on the postulates of the principle of relativity and on the experimental fact that the speed of propagation of light in vacuum is constant. When put together, these two postulates then led us to the mixing of the time and the spatial coordinates. As a consequence, we obtained the Lorentz transformation as a generalization of the nonrelativistic Galilean transformation. Thus events are expressed in terms of the time t and the three spatial coordinates x, y, z. Consequently, we are naturally led to the notion of a four-dimensional spacetime. Such a spacetime was indeed introduced by Minkowski in 1908 [1]. In this chapter the four-dimensional formulation of space and time is given [2-7].

5.1 Special relativity as a valuable guide

Before introducing the four-dimensional formulation of special relativity, it is worthwhile mentioning the following.

Special relativity is a kinematical rather than a dynamical theory. It is actually a skeleton theory and, as such, it provides a background and a guideline frame to the other dynamical theories of fields and matter. It imposes restrictions on the laws of physics which physical theories can have. It demands that every general law of physics, expressed in terms of the coordinates t, x, y, z of the system K, should be so that it can be transformed, by means of the Lorentz transformation, into a law of exactly the same form when expressed in terms of the new coordinates t', x', y', z' of the system K'. In other words, the laws of physics should be *invariant* under the Lorentz transformation. As a result, the theory is a valuable guide when looking for new laws of physics. A law of physics which cannot be written in a Lorentz-invariant form is simply not valid, or is only an approximate law.

5.2 Four dimensions in classical mechanics

It is well known that classical mechanics is based on a four-dimensional manifold of three-dimensional space and the time. However, there is an *essential* difference between the concepts of space and time in classical mechanics and the four-dimensional spacetime of special relativity.

In classical mechanics the three-dimensional subspace with constant t is absolute and is independent of the inertial coordinate system. This means one has a separate three-dimensional space, along with a one-dimensional time coordinate, i.e., $O(3) \times T(1)$.

In special relativity, on the other hand, the spatial and time coordinates appear in the laws of physics at exactly the same footing, i.e., O(1,3).

Without a four-dimensional formulation, one can also carry out a Lorentz transformation in order to check the invariance of a given law in special relativity. Indeed this was the case when Einstein first proved in his historical paper that Maxwell's equations are invariant under the Lorentz transformation. But such a procedure is quite lengthy, and it should be done for each and every field equation we have in physics. The four-dimensional formalism, on the other hand, provides a simple way to ensure Lorentz invariance by the form of the law itself.

5.3 The Minkowskian spacetime

If we use as coordinates of an event the quantities $x^0 = ct$, $x^1 = x$, $x^2 = y$, $x^3 = z$, then x^0, x^1, x^2, x^3 may be considered as the components of a vector in four dimensions. The four-dimensional space provided by these four-vectors is then called the *Minkowskian spacetime*. The square of the length of this four-vector,

$$\left(x^0\right)^2 - \left(x^1\right)^2 - \left(x^2\right)^2 - \left(x^3\right)^2 = c^2 t^2 - x^2 - y^2 - z^2, \quad (5.1)$$

does not change under "rotations" of the four-dimensional coordinate system, that is under the Lorentz transformation. If Λ is the 4×4 matrix of the Lorentz transformation whose elements are $\Lambda^\alpha{}_\beta$, then the transformed coordinates are given by

$$x'^\alpha = \Lambda^\alpha{}_\beta x^\beta, \quad (5.2)$$

where Greek indices take the values 0, 1, 2, 3. Invariance of the expression (5.1) then means that

$$\left(x'^0\right)^2 - \left(x'^1\right)^2 - \left(x'^2\right)^2 - \left(x'^3\right)^2 = \left(x^0\right)^2 - \left(x^1\right)^2 - \left(x^2\right)^2 - \left(x^3\right)^2, \tag{5.3}$$

where x'^α are given by Eq. (5.2).

In general a set of four quantities V^α, which transform like the components of the coordinates x^α under the Lorentz transformation, is called a *four-vector*. One can extend the definition of vectors to *tensors* of any order in the Minkowskian spacetime. The event described by the *position four-vector* x^α is just an example of such quantities. A *scalar* is then a tensor of order 0, whereas a vector is a tensor of order 1. Under a Lorentz transformation a scalar is left *invariant*, a four-vector V^α transforms like the coordinates,

$$V'^\alpha = \Lambda^\alpha{}_\beta V^\beta, \tag{5.4}$$

whereas a tensor $T^{\alpha\beta}$ of order 2, for example, transforms like a product of coordinates,

$$T'^{\alpha\beta} = \Lambda^\alpha{}_\gamma \Lambda^\beta{}_\delta T^{\gamma\delta}, \tag{5.5}$$

and so on.

As was discussed in previous sections, the invariance of any physical law under the Lorentz transformation is evident once it is expressed in a *covariant* four-dimensional form. All terms of the law should then be four-tensors of the same order. There is also the possibility of spinor formulation, in addition to the four-tensor formalism. The physical law should again be formulated covariantly. (For the theory of spinors and their relation to the Lorentz group, see [8-10].) A physical law which does not satisfy these requirements cannot be put in a covariant form. The four-dimensional transformation properties of the terms of a physical law, therefore, enable examining its relativistic validity. In the following some more details on four-vectors are given.

The Minkowskian spacetime

Four-vectors are natural generalization to the ordinary three-vectors of classical mechanics. However, use is made in the four-dimensional case of the flat-space metric

$$\eta_{\alpha\beta} = \begin{pmatrix} +1 & 0 & 0 & 0 \\ 0 & -1 & 0 & 0 \\ 0 & 0 & -1 & 0 \\ 0 & 0 & 0 & -1 \end{pmatrix} \tag{5.6}$$

and its inverse matrix $\eta^{\alpha\beta}$ (having the same expression as $\eta_{\alpha\beta}$), instead of the three-dimensional unit matrix. The metrics $\eta_{\alpha\beta}$ and $\eta^{\alpha\beta}$ can then be used to *lower* and *raise* the indices of four-quantities,

$$V_\alpha = \eta_{\alpha\beta} V^\beta, \quad V^\alpha = \eta^{\alpha\beta} V_\beta, \tag{5.7}$$

$$T_{\alpha\beta} = \eta_{\alpha\gamma} \eta_{\beta\delta} T^{\gamma\delta}, \quad T^{\alpha\beta} = \eta^{\alpha\gamma} \eta^{\beta\delta} T_{\gamma\delta}, \tag{5.8}$$

and so on for tensors of higher orders. Quantities with lower indices, like $T_{\alpha\beta}$, are called *covariant* whereas those with upper indices, such as $T^{\alpha\beta}$, are referred to as *contravariant*.

The *scalar product* of two four-vectors V^α and W^α is defined by

$$V_\alpha W^\alpha = \eta_{\alpha\beta} V^\beta W^\alpha = \eta^{\alpha\beta} V_\alpha W_\beta, \tag{5.9}$$

and it is a scalar (Lorentz invariant).

In analogy to the position four-vector x^α, the zeroth component of any four-vector is called *timelike* whereas the other three components are called *spacelike*. The square $V_\alpha V^\alpha$ of a four-vector V^α can be positive, zero, or negative. The four-vector is accordingly called *timelike*, *null*, or *spacelike*, respectively. A timelike vector is called *positive* or *negative* according to whether its timelike component is positive or negative, respectively. The manifold of all null vectors forms the *light cone* (a detailed discussion of which is given in Chapter 6).

The tensor δ_β^α is defined by

$$\delta_\beta^\alpha = \begin{cases} 1, & \alpha = \beta \\ 0, & \alpha \neq \beta \end{cases} \tag{5.10}$$

in all coordinate systems. It is called the *Kronecker delta*, and it satisfies

$$V_\alpha = \delta_\alpha^\beta V_\beta, \tag{5.11a}$$
$$V^\alpha = \delta_\beta^\alpha V^\beta \tag{5.11b}$$

for any vector V_α.

From any tensor $T_{\alpha\beta}$ of order 2 one can form the scalar

$$T_\alpha{}^\alpha = \delta_\beta^\alpha T_\alpha{}^\beta, \tag{5.12}$$

called the *trace* of the tensor.

A tensor is called symmetric with respect to two of its indices if their exchange does not affect the value of the tensor. Thus, for instance, the tensor $T_{\alpha\beta\gamma}$ of order 3 is symmetric with respect to the indices α and β if

$$T_{\beta\alpha\gamma} = T_{\alpha\beta\gamma}. \tag{5.13}$$

A tensor $A_{\alpha\beta\gamma}$ is called *antisymmetric* (or *skew-symmetric*) with respect to two of its indices α and β, for instance, if it satisfies

$$A_{\beta\alpha\gamma} = -A_{\alpha\beta\gamma}. \tag{5.14}$$

The diagonal components of an antisymmetric tensor $A_{\alpha\beta}$ of order 2, that is the components A_{00}, A_{11}, A_{22}, A_{33}, are all equal to zero since $A_{00} = -A_{00}$, and so on.

5.4 Proper time

When a particle moves in the ordinary three-dimensional space, it describes a path in the Minkowskian spacetime, called a *world*

Proper time

line. The four-vector dx^α represents the infinitesimal change in the position four-vector x^α, and it is a tangent vector to the world line.

The square of dx^α, namely $\eta_{\alpha\beta}dx^\alpha dx^\beta$, is a scalar, and it is denoted by ds^2. Accordingly we have

$$ds^2 = \eta_{\alpha\beta}dx^\alpha dx^\beta = c^2 dt^2 - dx^2 - dy^2 - dz^2. \tag{5.15}$$

The physical meaning of ds can best be understood if we evaluate it in an inertial coordinate system K' in which the particle is momentarily at rest. Denoting the coordinates in K' by x'^α, then in this system the timelike component of dx'^μ is $dx'^0 = cdt'$, whereas its spacelike components vanish, $dx'^k = 0$ ($k = 1, 2, 3$). Thus in the system K' one has

$$dx'^\alpha = (cdt', 0, 0, 0), \tag{5.16}$$

and consequently

$$ds^2 = c^2 dt^2 - dx^2 - dy^2 - dz^2 = c^2 dt'^2. \tag{5.17}$$

Accordingly, $d\tau = ds/c$ is the time interval as measured by a clock moving with the particle; it is called the *proper time*.

From Eq. (5.17) we obtain

$$d\tau = \frac{1}{c}\sqrt{c^2 dt^2 - dx^2 - dy^2 - dz^2}, \tag{5.18}$$

or

$$d\tau = dt\sqrt{1 - \frac{1}{c^2}\left[\left(\frac{dx}{dt}\right)^2 + \left(\frac{dy}{dt}\right)^2 + \left(\frac{dz}{dt}\right)^2\right]}. \tag{5.19}$$

But the expression in the square brackets on the right-hand side of Eq. (5.19) is simply v^2, where v is the velocity of the moving clock. Consequently we have

$$d\tau = dt\sqrt{1 - \frac{v^2}{c^2}}. \tag{5.20}$$

Equation (5.20) also follows from the time dilation formula given in Section 4.4, since $d\tau$ is the time interval of a clock moving with the particle and dt is the corresponding time interval as measured in a coordinate system from which the motion is observed.

The integration of Eq. (5.20) gives the time interval as measured by the moving clock,

$$\Delta \tau = \int_{t_1}^{t_2} \sqrt{1 - \frac{v^2}{c^2}} dt, \qquad (5.21)$$

whereas the corresponding time interval, as measured by a clock at rest, is given by $\Delta t = t_2 - t_1$.

Equation (5.21) shows that the proper time of a moving body is less than the corresponding time in the "rest" coordinate system; moving clocks go slower than those which are at rest.

Let us now have two inertial systems K and K', where K' is moving relative to K with the speed v. As viewed by an observer in the system K, the clocks in K' go slower than those in K. On the other hand, as judged by an observer in the coordinate system K', the clocks in K go slower than those in K'. It thus appears, at a first sight, that there is a contradiction. However, a careful analysis of the problem shows that there is no contradiction (see Problem 5.1).

A different problem, often referred to as the twins paradox, is that in which we have two clocks one of which goes along a *closed* path, returning to the starting point where the other clock is left behind at rest. Then clearly the moving clock must go slower than the one at rest. A converse reasoning, according to which the moving clock should be considered to be at rest and the other one as moving, is not possible; the clock going along the closed trajectory does not move with a constant velocity, and therefore a coordinate system that is attached to it cannot be inertial along the entire path. Indeed an experiment has been carried out with particles that decay, and the result predicted

Velocity and acceleration four-vectors

by special relativity was confirmed.

One can therefore conclude that the time interval shown by a clock is given by the integral

$$\int_a^b d\tau, \qquad (5.22)$$

where $d\tau = ds/c$ and the integration is carried out along the world line of the clock. For a clock at rest, the world line is a straight line parallel to the x^0 axis. If the clock goes along a closed path in the ordinary three-dimensional space, on the other hand, then its world line is a curve passing through the initial and the final points of the motion.

Figure 5.1 gives the Minkowskian spacetime diagram describing the world lines of two clocks one at rest, while the other moves along a closed curve, in the three-dimensional space. The world lines are described by the straight line L and the curve L' between the initial point a and the final point b of the motion. (The x^2 and x^3 axes are omitted for brevity.)

Finally, it will also be noted that the clock at rest always shows a longer time interval than that of the moving clock. In other words, the integral (5.22) becomes maximum if the integration is carried out along the straight world line connecting the points a and b.

5.5 Velocity and acceleration four-vectors

We continue our four-dimensional analysis by forming the velocity and acceleration four-vectors from their corresponding ordinary three-dimensional vectors.

The velocity four-vector of a particle is defined by

$$u^\alpha = \frac{dx^\alpha}{ds}, \qquad (5.23)$$

and its relation to the ordinary three-dimensional velocity **v** can be obtained by expressing ds in terms of dt, using Eq. (5.20). We then have

$$u^\alpha = \frac{1}{c\sqrt{1-\dfrac{v^2}{c^2}}}\frac{dx^\alpha}{dt}. \tag{5.24}$$

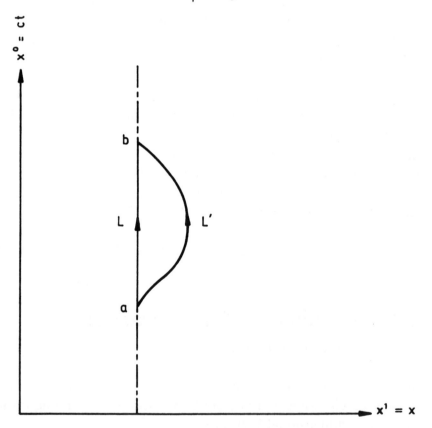

Figure 5.1 The Minkowskian spacetime diagram describing the world lines of two clocks one at rest, while the other moves along a closed curve, in the three-dimensional space.

Velocity and acceleration four-vectors

The timelike component of u^α is consequently given by

$$u^0 = \frac{1}{\sqrt{1 - \frac{v^2}{c^2}}}, \tag{5.25}$$

whereas its spacelike components

$$\mathbf{u} = \left(u^1,\ u^2,\ u^3\right) \tag{5.26}$$

are given by

$$\mathbf{u} = \frac{1}{\sqrt{1 - \frac{v^2}{c^2}}} \frac{\mathbf{v}}{c}. \tag{5.27}$$

Here

$$\mathbf{v} = (v_x,\ v_y,\ v_z) = \left(\frac{dx}{dt},\ \frac{dy}{dt},\ \frac{dz}{dt}\right) \tag{5.28}$$

is the ordinary three-dimensional velocity vector.

It will be noted that the velocity four-vector u^α is dimensionless. Moreover, by Eq. (5.15),

$$u_\alpha u^\alpha = 1, \tag{5.29}$$

i.e., its length is unity.

The acceleration four-vector of a particle is subsequently defined by

$$\frac{du^\alpha}{ds} = \frac{d^2 x^\alpha}{ds^2}, \tag{5.30}$$

which, by Eq. (5.29), satisfies the orthogonality condition

$$u_\alpha \frac{du^\alpha}{ds} = 0. \tag{5.31}$$

Using Eqs. (5.26), (5.27) and (5.29) we then find for the components of the acceleration four-vector the following:

$$\frac{du^0}{ds} = \frac{\gamma}{c} \frac{d\gamma}{dt}, \tag{5.32a}$$

$$\frac{d\mathbf{u}}{ds} = \frac{\gamma}{c^2}\frac{d(\gamma\mathbf{v})}{dt}, \qquad (5.32b)$$

where

$$\gamma = \frac{1}{\sqrt{1 - \frac{v^2}{c^2}}}. \qquad (5.33)$$

In the next chapter the light cone, an important feature of the four-dimensional spacetime structure, is discussed.

5.6 Problems

5.1 Given two inertial systems K and K', where the latter is moving relative to K with the speed v. As judged by an observer in K, the clocks in the system K' go slower than those in K. On the other hand, as viewed by an observer in K', the clocks in the system K go more slowly than those in K'. Show that there is really no contradiction between the above two observations.

Solution: The solution of the problem is left for the reader.

5.2 Find the timelike component of the acceleration four-vector of a particle in a coordinate system in which the particle is momentarily at rest. Express the result in terms of the ordinary velocity $\mathbf{v}$ and acceleration $\mathbf{a} = d\mathbf{v}/dt$ for the two limiting cases where once v changes only in direction and secondly v changes only in magnitude. Use Eqs. (5.32) to show that one obtains $\gamma^2|\mathbf{a}|$ and $\gamma^3|\mathbf{a}|$ for these limiting cases, where $\gamma = 1/\sqrt{1-\beta^2}$ and $\beta = v/c$ [4].

Solution: The solution is left for the reader.

5.7 References

[1] H. Minkowski, Space and time (an address delivered at the 80th Assembly of German Natural Scientists and Physicians, at

References

Cologne, Germany, 21 September, 1908); English translation in: *The Principle of Relativity* (Dover, New York, 1923), p. 73.

[2] A. Einstein, *Relativity: The Special and General Theory* (Crown Publishers, New York, 1931).

[3] A. Einstein, *The Meaning of Relativity* (Princeton University Press, Princeton, N.J., 1955).

[4] L.D. Landau and E.M. Lifshitz, *The Classical Theory of Fields* (Addison-Wesley, Reading, Massachusetts, 1959).

[5] A.P. French, *Special Relativity* (W.W. Norton, New York and London, 1968).

[6] A. Einstein, *Autobiographical Notes*, P.A. Schilpp, Editor (Open Court Publishing Company, La Salle and Chicago, Illinois, 1979).

[7] A. Einstein, *Ann. Physik* (Germany) **17**, 891 (1905); English translation in: A. Einstein *et al.: The Principle of Relativity* (Dover, New York, 1923).

[8] M. Carmeli, *Group Theory and General Relativity* (McGraw-Hill, New York, 1977).

[9] M. Carmeli, *Classical Fields: General Relativity and Gauge Theory* (John Wiley, New York, 1982).

[10] E. Cartan, *The Theory of Spinors* (The M.I.T. Press, Cambridge, Massachusetts, 1966).

Chapter 6

The Light Cone

In this chapter we discuss the light cone, an important description of the Minkowskian spacetime.

6.1 The light cone

In the four-dimensional spacetime, we choose an inertial coordinate system with coordinates $x^\alpha = (ct,\ x,\ y,\ z)$ and O describes the origin of the coordinate system. We now examine how other events are related to O [1-5].

A finite-mass particle moving with a constant velocity and passing through O will be represented by a straight line. The inclination of this straight line with respect to the x axis is such that the cotangent of its angle is equal to v/c, where v is the ordinary velocity. But the maximum of such a velocity is c. Hence the minimum angle is $\pi/4$. Accordingly, we have a cone which is represented by the formula

$$c^2 t^2 - \left(x^2 + y^2 + z^2\right) = 0, \tag{6.1}$$

called the *light cone*, whose symmetry axis coincides with the x^0 axis.

Figure 6.1 describes the light cone in two dimensions, one of which is the coordinate x^0 ($= ct$), whereas the other is the coordinate x^1 ($= x$). The propagation of two light signals in opposite directions passing through $x = 0$ at time $t = 0$, is represented by the two diagonal straight lines. The motion of finite-mass particles, on the other hand, are represented by straight lines in the interior of the light cone. (Compare the galaxy cone given in Figure 2.1 of Chapter 2.)

6.2 Events and coordinate systems

Let now two points in the four-dimensional Minkowskian space-time be given by

$$\begin{aligned} x_1^\alpha &= (ct_1,\ x_1,\ y_1,\ z_1), \\ x_2^\alpha &= (ct_2,\ x_2,\ y_2,\ z_2), \end{aligned} \quad (6.2)$$

in an inertial coordinate system K, and by

$$\begin{aligned} x_1'^\alpha &= (ct_1',\ x_1',\ y_1',\ z_1'), \\ x_2'^\alpha &= (ct_2',\ x_2',\ y_2',\ z_2'), \end{aligned} \quad (6.3)$$

in another system K'.

The *interval four-vectors* between the two points are defined by

$$\begin{aligned} X^\alpha &= (cT,\ X,\ Y,\ Z) \\ &= [c(t_2 - t_1),\ (x_2 - x_1),\ (y_2 - y_1),\ (z_2 - z_1)], \end{aligned} \quad (6.4a)$$

and

$$\begin{aligned} X'^\alpha &= (cT',\ X',\ Y',\ Z') \\ &= [c(t_2' - t_1'),\ (x_2' - x_1'),\ (y_2' - y_1'),\ (z_2' - z_1')], \end{aligned} \quad (6.4b)$$

Events and coordinate systems

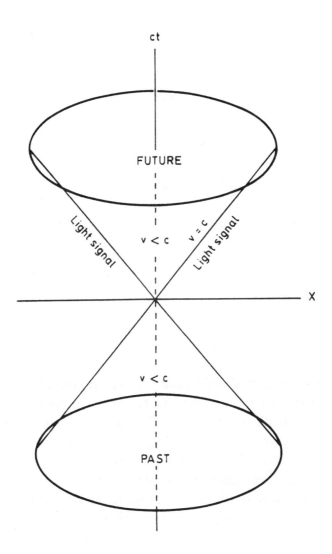

Figure 6.1 The light cone in two dimensions, $x^0 (= ct)$ and $x^1 (= x)$. The propagation of two light signals in opposite directions passing through $x = 0$ at time $t = 0$, is represented by the two diagonal straight lines. The motion of finite-mass particles, on the other hand, are represented by straight lines in the interior of the light cone. (Compare the galaxy cone given in Figure 2.1 of Chapter 2.)

in the two coordinate systems K and K', respectively. The squares of the intervals between the two events in the systems K and K' are then given by

$$X_\alpha X^\alpha = c^2 T^2 - R^2,$$
$$X'_\alpha X'^\alpha = c^2 T'^2 - R'^2, \qquad (6.5)$$

where

$$R^2 = X^2 + Y^2 + Z^2,$$
$$R'^2 = X'^2 + Y'^2 + Z'^2 \qquad (6.6)$$

are the three-dimensional distances in K and K'. Lorentz invariance then requires that

$$X_\alpha X^\alpha = c^2 T^2 - R^2 = c^2 T'^2 - R'^2 = X'_\alpha X'^\alpha. \qquad (6.7)$$

6.3 Problems

6.1 Show that if the interval four-vector X^α between two events is timelike in an inertial coordinate system K, then there exists a coordinate system K' in which the two events occur at the same spatial point at different times. The time difference of the two events in K' is given by $T' = \sqrt{X_\alpha X^\alpha}/c$. Show also that if the interval four-vector X^α between two events in the system K is spacelike, then there exists a coordinate system K' in which the two events occur simultaneously at separate spatial points with the distance $R' = -\sqrt{X_\alpha X^\alpha}$ between them.

Solution: The solution is left for the reader.

6.4 Future and past

Consider now the events within the light cone (see Figure 6.1) in which $c^2 t^2 - x^2 > 0$ and therefore the interval four-vector X^α is timelike. The upper part of the cone is called the *future* since all the events in it occur after O, and there is no coordinate system in which events can occur before O.

A similar analysis shows that all the events in the lower part of the light cone occur before O in all coordinate systems. This part is consequently called the *past*. For any event outside the light cone there exist coordinate systems in which the event occurs after and before O, and one system in which it occurs simultaneously with O.

In the next chapter the relationship between mass, energy and momentum will be given.

6.5 References

[1] H. Minkowski, Space and time (an address delivered at the 80th Assembly of German Natural Scientists and Physicians, at Cologne, 21 September, 1908); English translation in: *The Principle of Relativity* (Dover, New York, 1923), p.73.
[2] L.D. Landau and E.M. Lifshitz, *The Classical Theory of Fields* (Addison-Wesley, Reading, Massachusetts, 1959).
[3] D. Bohm, *The Special Theory of Relativity* (Benjamin, New York, 1965).
[4] M. Born, *Einstein's Theory of Relativity* (Dover, New York, 1962).
[5] A.P. French, *Special Relativity* (W.W. Norton, New York and London, 1968).

Chapter 7

Mass, Energy and Momentum

We conclude the fundamentals of the special relativity theory by discussing the *dynamical* concepts of mass, energy and momentum. It follows that one of the most important consequences of the special theory of relativity is the relationship between the mass and energy, stated in the form of a simple and universal law, $E = mc^2$. Before the advent of special relativity, there were two separate conservation laws of great importance. These were the conservation laws of mass and of energy, and they were independent of each other. The special theory of relativity united them into one law. In this chapter the relationship between the mass and energy is presented. This is done by introducing the momentum four-vector whose timelike component is the energy whereas its spatial components are those of the ordinary momentum.

7.1 Preliminaries

We start our discussion by presenting some physical comments about the relationship between the mass and energy [1-4].

According to the theory of special relativity, the total energy of a particle is not given by the familiar Newtonian kinetic energy $mv^2/2$, where m is the mass and v is the velocity of the particle. Rather, it is given by

$$\frac{mc^2}{\sqrt{1-\frac{v^2}{c^2}}}, \qquad (7.1)$$

where c is the speed of light.

Expanding (7.1) into a power series in v/c, we obtain

$$\frac{mc^2}{\sqrt{1-\frac{v^2}{c^2}}} = mc^2 + \frac{1}{2}mv^2 + \frac{3}{8}m\frac{v^4}{c^2} + \cdots. \qquad (7.2)$$

The second term on the right-hand side of Eq. (7.2) is, of course, the Newtonian kinetic energy. The first term, mc^2, is called the *rest energy* of the particle.

7.2 Mass, energy and momentum

We now derive the relationship between the mass, energy and momentum [1-4].

Starting with the relation between the energy and momentum of the photon,

$$E = cp, \qquad (7.3)$$

one can associate the inertial mass

$$m = \frac{E}{c^2} \qquad (7.4)$$

Mass, energy and momentum

to the photon. Hence, the inertial mass of a photon with the energy E is given by

$$m = \frac{p}{c}. \tag{7.5}$$

Equation (7.5) is a particular case of the formula

$$m = \frac{p}{v} \tag{7.6}$$

for a particle, familiar from the Newtonian mechanics, for $v = c$.

From Eqs. (7.4) and (7.6) we obtain

$$\frac{v}{c} = \frac{cp}{E}, \tag{7.7}$$

expressing the relationship between the energy, momentum and velocity.

From classical mechanics we know that the increase of kinetic energy due to work done by an external force f is given by

$$dE = f dx = \frac{dp}{dt} dx = v dp. \tag{7.8}$$

Equations (7.7) and (7.8) then yield

$$E dE = c^2 p dp, \tag{7.9}$$

the integration of which gives

$$E^2 - c^2 p^2 = E_0^2, \tag{7.10}$$

where E_0^2 is a constant of integration. Equation (7.10) is invariant under the Lorentz transformation as will be shown in the sequel.

To determine the constant E_0 we use Eqs. (7.7) and (7.10), getting

$$E = \frac{E_0}{\sqrt{1 - \frac{v^2}{c^2}}}. \tag{7.11}$$

Expanding now the right-hand side of this equation into a power series in v/c then gives

$$E = E_0 + \frac{1}{2}\left(\frac{E_0}{c^2}\right)v^2 + \frac{3}{8}\left(\frac{E_0}{c^2}\right)\frac{v^4}{c^2} + \cdots. \qquad (7.12)$$

This should now be compared to Eq. (7.2), giving

$$\frac{E_0}{c^2} = m_0. \qquad (7.13)$$

Here m_0 is a constant, called the *rest mass* of the particle (previously denoted by m).

Using Eq. (7.13) in Eqs. (7.10) and (7.11) then gives

$$E^2 - c^2 p^2 = m_0^2 c^4 \qquad (7.14)$$

and

$$E = \frac{m_0 c^2}{\sqrt{1 - \dfrac{v^2}{c^2}}}, \qquad (7.15)$$

respectively. Equation (7.4) can now be written in the form

$$m = \frac{m_0}{\sqrt{1 - \dfrac{v^2}{c^2}}}, \qquad (7.16)$$

where m is the inertial mass of the particle. The rest mass m_0 is, consequently, equal to the inertial mass for $v = 0$.

From Eqs. (7.15) and (7.16) one obtains

$$E = mc^2. \qquad (7.17)$$

This is Einstein's famous formula expressing the total energy of the particle in terms of its inertial mass. Moreover, the three-dimensional momentum **p** is now given by

$$\mathbf{p} = m\mathbf{v}, \qquad (7.18)$$

where $\mathbf{v} = d\mathbf{x}/dt$ is the three-dimensional velocity and m is the inertial mass rather than the rest mass m_0.

It should be emphasized that none of the equations (7.11) and (7.14)–(7.18) demands that the velocity v of the particle should be constant even though they were all derived from special relativity, a theory based on transformations between inertial systems moving relative to each other with constant velocities. In fact, these formulas are valid for nonconstant velocities v as well.

Physical phenomena described within the framework of special relativity are by no means restricted to processes with constant velocities. The velocity v can be replaced by any other variable related to it, such as the momentum p or the angular momentum J, for instance. All one needs is expressing the ratio $\beta = v/c$ in the Lorentz contraction factor $\sqrt{1-\beta^2}$, appearing in the above equations of the particle, by its value in terms of the desired new variable.

Let us, for instance, express the Lorentz contraction factor as a function of the momentum p. Using Eq. (7.18), we then obtain

$$\frac{1}{1-\dfrac{v^2}{c^2}} = 1 + \frac{p^2}{p_0^2}, \qquad (7.19)$$

where

$$p_0 = m_0 c = \frac{E_0}{c} \qquad (7.20)$$

is a characteristic constant of the momentum of the particle. As v takes the values $0 \leq v < c$, the momentum goes from zero to infinity, $0 \leq p < \infty$, and consequently the expressions on both sides of Eq. (7.19) take the values 1 to ∞, as expected. Notice that for a photon $p_0 = 0$ since its rest mass m_0 equals to zero.

Equations (7.15) and (7.16) will consequently assume the

forms

$$E = E_0\sqrt{1+\frac{p^2}{p_0^2}} = \frac{p_0^2}{m_0}\sqrt{1+\frac{p^2}{p_0^2}} \qquad (7.21)$$

and

$$m = m_0\sqrt{1+\frac{p^2}{p_0^2}}, \qquad (7.22)$$

respectively, when the new expression of the Lorentz contraction factor is used. Notice that Eq. (7.21) is completely equivalent to the energy-momentum formula (7.14).

Expanding now Eq. (7.21) in powers of p/p_0, we then obtain for the energy of the particle as a function of its momentum,

$$E(p) = E_0\left(1+\frac{1}{2}\frac{p^2}{p_0^2}-\frac{1}{8}\frac{p^4}{p_0^4}+\frac{1}{16}\frac{p^6}{p_0^6}+\cdots\right). \qquad (7.23)$$

This should then be compared with its equivalent expression in terms of the velocity,

$$E(v) = E_0\left(1+\frac{1}{2}\frac{v^2}{c^2}+\frac{3}{8}\frac{v^4}{c^4}+\frac{5}{16}\frac{v^6}{c^6}+\cdots\right), \qquad (7.24)$$

obtained by expanding Eq. (7.15) in powers of v/c.

The first terms on the right-hand sides of Eqs. (7.23) and (7.24) are equal to the rest energy of the particle. However, the second term on the right-hand side of Eq. (7.23) is not equal to the second term of Eq. (7.24) since, by Eq. (7.18), p is proportional to m rather than to m_0. The third terms on the right-hand sides of Eqs. (7.23) and (7.24) appear with opposite signs.

7.3 Angular-momentum representation

Likewise, one can make use of the angular momentum J of the particle by expressing the ratio p^2/p_0^2 in terms of J [5]. Using

Angular-momentum representation

Eq. (7.20) one then finds

$$\frac{p^2}{p_0^2} = \frac{p^2}{m_0^2 c^2} = \frac{1}{m_0 c^2} \frac{p^2}{m_0} = \frac{1}{m_0 c^2} \frac{J^2}{I_0}, \quad (7.25)$$

where I_0 is the *rest moment of inertia* of the particle (calculated, using the rest mass). The last equality in the above equation is a result of expressing the kinetic energy in terms of p and J. Accordingly one obtains

$$\frac{p^2}{p_0^2} = \frac{J^2}{J_0^2}, \quad (7.26)$$

where

$$J_0 = \sqrt{m_0 c^2 I_0} \quad (7.27)$$

is a characteristic constant of the angular momentum of the particle. Notice that for a photon $J_0 = 0$, since both m_0 and I_0 are equal to zero in this case.

Equation (7.19) then gives

$$\frac{1}{1 - \frac{v^2}{c^2}} = 1 + \frac{J^2}{J_0^2}, \quad (7.28)$$

and, as a consequence, Eqs. (7.21)–(7.23) give

$$E = E_0 \sqrt{1 + \frac{J^2}{J_0^2}} = \frac{J_0^2}{I_0} \sqrt{1 + \frac{J^2}{J_0^2}}, \quad (7.29)$$

$$m = m_0 \sqrt{1 + \frac{J^2}{J_0^2}}, \quad (7.30)$$

$$E(J) = E_0 \left(1 + \frac{1}{2}\frac{J^2}{J_0^2} - \frac{1}{8}\frac{J^4}{J_0^4} + \frac{1}{16}\frac{J^6}{J_0^6} + \cdots \right). \quad (7.31)$$

Finally, just as Eq. (7.21) is completely equivalent to the energy-momentum formula (7.14), we can rewrite Eq. (7.29) as an energy-angular-momentum formula. We then have

$$E^2 - \gamma^2 J^2 = I_0^2 \gamma^4, \qquad (7.32)$$

where

$$\gamma = c\sqrt{\frac{m_0}{I_0}} \qquad (7.33)$$

is a natural angular velocity of the particle, and the expression on the right-hand side of Eq. (7.32) is equal to $m_0^2 c^4$. The equivalent formulas to Eqs. (7.15)–(7.18) and (7.7) will then have the forms:

$$E = \frac{I_0 \gamma^2}{\sqrt{1 - \frac{\Omega^2}{\gamma^2}}}, \qquad (7.34)$$

$$I = \frac{I_0}{\sqrt{1 - \frac{\Omega^2}{\gamma^2}}}, \qquad (7.35)$$

$$E = I\gamma^2, \qquad (7.36)$$

$$\mathbf{J} = I\mathbf{\Omega}, \qquad (7.37)$$

$$\frac{\Omega}{\gamma} = \frac{\gamma J}{E}, \qquad (7.38)$$

where Ω is the angular velocity of the particle satisfying the condition $0 \leq \Omega < \gamma$. We may also add the relation

$$\frac{1}{1 - \frac{\Omega^2}{\gamma^2}} = 1 + \frac{J^2}{J_0^2}, \qquad (7.39)$$

which is equivalent to Eq. (7.19).

7.4 Energy-momentum four-vector

To conclude this chapter we introduce the *energy-momentum four-vector* defined by

$$p^\alpha = m_0 c^2 u^\alpha, \qquad (7.40)$$

where m_0 is the rest mass of the particle, and u^α is the velocity four-vector defined by Eq. (5.23),

$$u^\alpha = \frac{dx^\alpha}{ds}. \qquad (7.41)$$

Using Eqs. (5.24)–(5.28) one then finds that the components of p^α are given by

$$p^0 = m_0 c^2 u^0 = \frac{m_0 c^2}{\sqrt{1-\beta^2}} = mc^2 = E, \qquad (7.42a)$$

$$p^k = m_0 c^2 u^k = \frac{m_0 c}{\sqrt{1-\beta^2}} \frac{dx^k}{dt} = mc \frac{dx^k}{dt}, \qquad (7.42b)$$

for $k = 1, 2, 3$. Thus the energy-momentum four-vector is given by

$$p^\alpha = \left(p^0, \, p^k\right) = (E, \, c\mathbf{p}), \qquad (7.43a)$$

$$p_\alpha = (p_0, \, p_k) = (E, \, -c\mathbf{p}), \qquad (7.43b)$$

where $\mathbf{p} = m d\mathbf{x}/dt$ is the ordinary three-dimensional momentum given by Eq. (7.18).

The square of p^α is consequently given by

$$p_\alpha p^\alpha = m_0^2 c^4 u_\alpha u^\alpha. \qquad (7.44)$$

Thus using Eqs. (5.29) and (7.43), we obtain

$$E^2 - c^2 \mathbf{p}^2 = m_0^2 c^4, \qquad (7.45)$$

that is, the relation (7.14) between mass, energy and momentum.

With the above discussion on mass, energy and momenta we end this part of the book on the fundamentals of the special theory of relativity.

7.5 Problems

7.1 Assume that $J_0 = \hbar$ for particles with *intrinsic* angular momentum (spin), such as electrons, where $\hbar = h/2\pi$ and h is Planck's constant. Use Eq. (7.27) to calculate the tangential velocity of a material point in such bodies due to their internal rotation [5].

Solution: The tangential velocity of a material point in such a body is defined by

$$v_g = \omega_0 r_g, \qquad (1)$$

where ω_0 is the angular velocity of the internal rotation of the body, and r_g, defined by

$$I_0 = m_0 r_g^2, \qquad (2)$$

is its *radius of gyration*.

Equation (7.27) with $J_0 = \hbar$, along with the ordinary relationship among angular momentum, moment of inertia and angular velocity, then give

$$J_0 = \sqrt{m_0 c^2 I_0} = I_0 \omega_0 = \hbar. \qquad (3)$$

Using now Eqs. (1)–(3) one then obtains

$$v_g = \omega_0 r_g = \frac{\hbar}{I_0}\sqrt{\frac{I_0}{m_0}} = \frac{\hbar}{\sqrt{m_0 I_0}} = \frac{\hbar c}{J_0} = c, \qquad (4)$$

that is, the speed of light.

7.2 Use the special relativistic line element

$$c^2 dt^2 - \left(dx^2 + dy^2 + dz^2\right) = ds^2 \qquad (1)$$

to derive the relationship between the energy and the momentum.

Solution: A straightforward calculation, using Eq. (1), gives

$$c^2 dt^2 \left(1 - \frac{v^2}{c^2}\right) = ds^2, \tag{2}$$

thus

$$\frac{dt}{d\tau} = \frac{1}{\sqrt{1 - \frac{v^2}{c^2}}}, \tag{3}$$

where $d\tau$ is related to ds by $ds = cd\tau$.

Multiplying now Eq. (2) by $m_0^2 c^4 / ds^2$, and using Eq. (3), one then obtains

$$\left(1 - \frac{v^2}{c^2}\right)^{-1} \left(m_0^2 c^4 - m_0^2 c^2 v^2\right) = m_0^2 c^4. \tag{4}$$

Using now the relationship between the inertial mass and the rest mass,

$$m = \frac{m_0}{\sqrt{1 - \frac{v^2}{c^2}}}, \tag{5}$$

in Eq. (4), we obtain

$$m^2 c^4 - c^2 \mathbf{p}^2 = m_0^2 c^4, \tag{6}$$

where $\mathbf{p} = m\mathbf{v}$.

7.6 References

[1] A. Einstein, *Relativity: The Special and General Theory* (Crown Publishers, New York, 1931).
[2] A.P. French, *Special Relativity* (W.W. Norton, New York and London, 1968).
[3] D. Bohm, *The Special Theory of Relativity* (Benjamin, New

York, 1965).

[4] M. Born, *Einstein's Theory of Relativity* (Dover, New York, 1962).

[5] M. Carmeli, *Nuovo Cimento Letters* **37**, 205 (1983).

Chapter 8

Velocity, Acceleration and Cosmic Distances

In this chapter we conclude the fundamentals of the cosmological special relativity (CSR) by discussing the *dynamical* concepts of *velocity, acceleration* and *cosmic distances* in spacevelocity. These concepts occur in CSR just as those of mass, linear momentum and energy appear in Einstein's special relativity (ESR) of spacetime (see Chapter 7).

8.1 Preliminaries

The most important result of Einstein's special relativity is probably the relationship between mass and energy (see Chapter 7). How it happened that the mass became so critical in this theory? The answer is very simple. The theory involves the square of the speed of light, c^2. What physical quantity incorporates the square of velocity? It is the energy, $mv^2/2$. Hence it is the mass that goes with v^2 and c^2. Thus m is taken as an invariant under the Lorentz transformation. This becomes the rest mass m_0. The inertial mass m follows to depend on the velocity.

What is the comparable physical quantity in cosmological special relativity? Certainly not the mass. Here we have τ^2, the square of the Hubble time constant. What physical quantity goes with the square of time? It is the acceleration **a** because $\mathbf{a}t^2/2$ describes the distance a particle makes at time t when it is subject to acceleration **a**. Hence **a** should be taken as the invariant quantity under the cosmological transformation. But then the acceleration depends on the cosmic time just as the mass depends on the velocity.

In this chapter these concepts are explored.

8.2 Velocity and acceleration four-vectors

We start our four-dimensional spacevelocity analysis by defining the velocity and acceleration (see Section 5.5 for the special relativistic treatment in spacetime).

The *velocity* four-vector of a particle in spacevelocity is defined as the dimensionless quantity

$$u^\mu = \frac{dx^\mu}{ds}, \qquad (8.1)$$

where $\mu = 0, 1, 2, 3$, $x^\mu = (x^0, x^1, x^2, x^3) = (\tau v, x, y, z)$, and τ is Hubble's time in the limit of zero gravity, a universal constant whose value is 12.486 Gyr [1,2]. In flat spacevelocity one has for the line element,

$$\tau^2 dv^2 - \left(dx^2 + dy^2 + dz^2\right) = ds^2, \qquad (8.2)$$

thus

$$\tau^2 \left(\frac{dv}{ds}\right)^2 \left(1 - \frac{dx^2 + dy^2 + dz^2}{\tau^2 dv^2}\right) = 1. \qquad (8.3)$$

Velocity and acceleration four-vectors

This gives

$$\tau^2 \left(\frac{dv}{ds}\right)^2 \left(1 - \frac{t^2}{\tau^2}\right) = 1, \tag{8.4}$$

and therefore

$$\frac{dv}{ds} = \frac{1}{\tau\sqrt{1 - \frac{t^2}{\tau^2}}}. \tag{8.5}$$

The *velocity* four-vector in spacevelocity can thus be expressed as

$$u^\mu = \frac{dx^\mu}{ds} = \frac{dx^\mu}{dv}\frac{dv}{ds} = \frac{1}{\tau\sqrt{1 - \frac{t^2}{\tau^2}}}\frac{dx^\mu}{dv}. \tag{8.6}$$

The velocitylike component of u^μ is therefore given by

$$u^0 = \gamma, \tag{8.7}$$

whereas its spatial components

$$\mathbf{u} = u^k = \left(u^1, u^2, u^3\right) \tag{8.8}$$

are given by

$$u^k = \frac{\gamma}{\tau}\frac{dx^k}{dv}, \tag{8.9}$$

where

$$\gamma = \frac{1}{\sqrt{1 - \frac{t^2}{\tau^2}}}. \tag{8.10}$$

It will be noted that, by Eq. (8.1),

$$u_\alpha u^\alpha = 1, \tag{8.11}$$

namely, the length of u^μ is unity.

The *acceleration* four-vector of a particle in spacevelocity is defined by
$$\frac{du^\mu}{ds} = \frac{d^2x^\mu}{ds^2}. \tag{8.12}$$

By differentiating Eq. (8.11) we find that the acceleration four-vector satisfies the orthogonality condition
$$u_\alpha \frac{du^\alpha}{ds} = 0. \tag{8.13}$$

The components of the acceleration four-vector of a particle in spacevelocity are, by Eqs. (8.5) and (8.7)–(8.9), then given by
$$\frac{du^0}{ds} = \frac{du^0}{dv}\frac{dv}{ds} = \frac{\gamma}{\tau}\frac{d\gamma}{dv}, \tag{8.14}$$

$$\frac{du^k}{ds} = \frac{du^k}{dv}\frac{dv}{ds} = \frac{\gamma}{\tau^2}\frac{d}{dv}\left(\gamma\frac{dx^k}{dv}\right), \tag{8.15}$$

where γ is given by Eq. (8.10).

8.3 Acceleration and distances

Multiplying Eq. (A.4) by $\mathbf{a}_0^2\tau^4$, where $\mathbf{a}_0$ is the ordinary three-vector acceleration as measured in the cosmic frame of reference (see Section 2.5) at cosmic time $t = 0$ (i.e. now). We obtain
$$\mathbf{a}_0^2\tau^6\left(\frac{dv}{ds}\right)^2\left(1 - \frac{t^2}{\tau^2}\right) = \mathbf{a}_0^2\tau^4. \tag{8.16}$$

Using now Eq. (8.5) in Eq. (8.16) the latter then yields
$$\left(1 - \frac{t^2}{\tau^2}\right)^{-1}\left(\mathbf{a}_0^2\tau^4 - \mathbf{a}_0^2 t^2\tau^2\right) = \mathbf{a}_0^2\tau^4. \tag{8.17}$$

Acceleration and distances

We now define the acceleration **a** at an arbitrary cosmic time t by

$$\mathbf{a} = \frac{\mathbf{a}_0}{\sqrt{1 - \frac{t^2}{\tau^2}}}, \tag{8.18}$$

then Eq. (8.17) will have the form

$$\mathbf{a}^2\tau^4 - \mathbf{a}^2 t^2 \tau^2 = \mathbf{a}_0^2 \tau^4, \tag{8.19}$$

or

$$\mathbf{a}^2\tau^4 - \tau^2 \mathbf{v}^2 = \mathbf{a}_0^2 \tau^4, \tag{8.20}$$

where $\mathbf{v} = \mathbf{a}t$ is the ordinary three-dimensional velocity. Equation (8.20) is the analog to

$$m^2 c^4 - c^2 \mathbf{p}^2 = m_0^2 c^4, \tag{8.21}$$

in ESR, where $\mathbf{p} = m\mathbf{v}$ is the linear momentum (see Section 7.2). Thus **a** reduces to $\mathbf{a}_0$ when the cosmic time $t = 0$ (i.e. at present).

The comparable to Eq. (8.18) in ESR is, of course,

$$m = \frac{m_0}{\sqrt{1 - \frac{v^2}{c^2}}}, \tag{8.22}$$

where m and m_0 are the inertial mass and the rest mass of the particle, with m reduces to m_0 at $v = 0$. If we multiply Eq. (8.22) by c^2 and expand both sides in v/c, we obtain

$$mc^2 = m_0 c^2 + \frac{m_0}{2} v^2 + \cdots. \tag{8.23}$$

Doing the same with Eq. (8.18) but multiplication by τ^2, and expanding in t/τ, we obtain

$$\mathbf{a}\tau^2 = \mathbf{a}_0 \tau^2 + \frac{\mathbf{a}_0}{2} t^2 + \cdots. \tag{8.24}$$

Equation (8.24) in CSR is of course the analog to Eq. (8.23) in ESR.

8.4 Energy in ESR versus cosmic distance in CSR

While Eq. (8.23) yields

$$E = E_0 + \frac{m_0}{2}v^2 + \cdots, \tag{8.25}$$

Eq. (8.24) gives

$$\mathbf{S} = \mathbf{S}_0 + \frac{\mathbf{a}_0}{2}t^2 + \cdots. \tag{8.26}$$

In the above equations E is, of course, the energy of the particle, whereas $\mathbf{S}$ is the cosmic distance. $E_0 = m_0 c^2$ is the rest energy of the particle whereas $m_0 v^2/2$ is the Newtonian kinetic energy. What about the terms in Eq. (8.26)? The term $\mathbf{a}_0 t^2/2$ is, of course, the Newtonian distance the particle makes due to the acceleration $\mathbf{a}_0$, the term $\mathbf{S}_0 = \mathbf{a}_0 \tau^2$ is unique to CSR, and it might be called the *intrinsic* cosmic distance of the particle.

Equation (8.20) can now be written as

$$\mathbf{S}^2 - \tau^2 \mathbf{v}^2 = \mathbf{S}_0^2, \tag{8.27}$$

in complete analogy to

$$E^2 - c^2 \mathbf{p}^2 = E_0^2 \tag{8.28}$$

in ESR.

8.5 Distance-velocity four-vector

We now define the cosmic distance-velocity four-vector. It is the analogous to the energy-momentum four-vector in ESR (see Section 7.4). It is defined by

$$\mathbf{v}^\mu = \mathbf{a}_0 \tau^2 u^\mu, \tag{8.29}$$

Distance-velocity four-vector

where u^μ has been defined in Section 8.2. We have

$$v^0 = a_0\tau^2 u^0, \tag{8.30a}$$

$$v^k = a_0\tau^2 u^k, \tag{8.30b}$$

where u^0 and u^k are given by Eqs. (8.7)–(8.10), with

$$v_0 = v^0, \quad v_k = -v^k. \tag{8.31}$$

Accordingly we have

$$v^2 = v_0 \cdot v^0 + v_k \cdot v^k = a_0^2\tau^4\left(u_0 u^0 + u_k u^k\right)$$

$$= a_0^2\tau^4 u_\alpha u^\alpha = a_0^2\tau^4 = S_0^2. \tag{8.32}$$

But, using Eqs. (8.7), (8.9) and (8.18),

$$v^0 = a_0\tau^2\gamma = a\tau^2 = S, \tag{8.33a}$$

$$v^k = a_0\tau\gamma\frac{dx^k}{dv} = a\tau\frac{dx^k}{dv} = a\tau\frac{dx^k}{dt}\frac{dt}{dv} = \tau\mathbf{v}, \tag{8.33b}$$

where $\mathbf{v}$ is the three-dimensional velocity. Hence

$$v_\alpha \cdot v^\alpha = v_0 \cdot v^0 + v_k \cdot v^k = a_0^2\tau^4\gamma^2 - a_0^2\tau^2\gamma^2\left(\frac{dx^k}{dv}\right)^2$$

$$= a^2\tau^4 - \tau^2 \mathbf{v}^2 = S^2 - \tau^2\mathbf{v}^2, \tag{8.34}$$

and accordingly, using (8.32),

$$S^2 - \tau^2\mathbf{v}^2 = S_0^2, \tag{8.35}$$

which is exactly Eq. (8.20) with $S = a\tau^2$ and $S_0 = a_0\tau^2$. The above analysis also shows that Eq. (8.35) is covariant under spacevelocity transformation.

Equation (8.35) is the analog to

$$E^2 - c^2\mathbf{p}^2 = E_0^2 \tag{8.36}$$

in ESR, where E and E_0 are the energy and rest energy, respectively, $\mathbf{S}$ and $\mathbf{S}_0$ are the cosmic distances at cosmic time t and present time $(t=0)$, respectively.

Finally, in ESR when the rest mass is zero (like the photon), one then has

$$E = cp, \tag{8.37}$$

which is valid at the light cone (see Chapter 6). In the case of CSR one also has, when $\mathbf{S}_0 = 0$,

$$\mathbf{S} = \tau\mathbf{v}, \tag{8.38}$$

now valid at the galaxy cone (see Section 2.14).

8.6 Conclusions

In this chapter it has been shown that the comparable quantity to the mass in Einstein's special relativity is the ordinary acceleration three-vector in cosmological special relativity. They both have similar behavior, one with respect to v/c and the other with t/τ:

$$m = \frac{m_0}{\sqrt{1 - \dfrac{v^2}{c^2}}},$$

and

$$\mathbf{a} = \frac{\mathbf{a}_0}{\sqrt{1 - \dfrac{t^2}{\tau^2}}}.$$

Furthermore, the role of the energy in Einstein's theory is being taken by the cosmic distance,

$$E = mc^2,$$

and
$$\mathbf{S} = \mathbf{a}\tau^2.$$

Finally, the analog of the energy formula
$$E^2 - c^2\mathbf{p}^2 = E_0^2$$
in ordinary special relativity is
$$\mathbf{S}^2 - \tau^2\mathbf{v}^2 = \mathbf{S}_0^2$$
in cosmological special relativity.

8.7 References

[1] M. Carmeli and T. Kuzmenko, Value of the cosmological constant: theory versus experiment, in: *Proceedings of the 20th Texas Symposium on Relativistic Astrophysics*, held 10-15 December 2000, Austin, Texas, J.C. Wheeler and H. Martel, Editors (American Institute of Physics, 2001).
[2] M. Carmeli, Accelerating universe, cosmological constant and dark energy, to be published.

Chapter 9

First Days of the Universe

The early stage of the Universe is discussed and the time lengths of its first days are given. If we denote the Hubble time in the zero-gravity limit by τ (approximately 12.5 Gyr), and T_n denotes the length of the n-th day, then we have the very simple relation $T_n = \tau/(2n-1)$. Hence we obtain for the first days the following lengths of time: $T_1 = \tau$, $T_2 = \tau/3$, $T_3 = \tau/5$, etc. [1].

9.1 Preliminaries

In this Chapter the lengths of days since the early Universe are calculated, day by day, from the first day after the Big Bang up to our present time. We find that the first day actually lasted the Hubble time in the limit of zero gravity, τ, which equals about 12.49 billion years. If T_n denotes the length of the n-th day in units of times of the early Universe, then we have a very simple relation

$$T_n = \frac{\tau}{2n-1}. \tag{9.1}$$

109

Hence we obtain for the first few days the following lengths of time:

$$T_1 = \tau, \; T_2 = \frac{\tau}{3}, \; T_3 = \frac{\tau}{5}, \; T_4 = \frac{\tau}{7}, \; T_5 = \frac{\tau}{9}, \; T_6 = \frac{\tau}{11}. \quad (9.2)$$

It also follows that the accumulation of time from the first day to the second, third, fourth, ..., up to now is just exactly the Hubble time. The Hubble time in the limit of zero gravity is the maximum time allowed in nature.

9.2 Lengths of days

Using Cosmological Special Relativity, the calculation is very simple. We assume that the Big Bang time with respect to us now was $t_0 = \tau$, the time of the first day after that was t_1, the time of the second day was t_2, and so on. In this way the time scale is progressing in units of one day (24 hours) in our units of present time. The time difference between t_0 and t_1, denoted by T_1, is the time as measured at the early Universe and is by no means equal to one day of our time. In this way we denote the times elapsed from the Big Bang to the end of the first day t_1 by T_1, between the first day t_1 and the second day t_2 by T_2 and so on. According to the rule of the addition of cosmic times one has (see Chapter 2), for example,

$$t_6 + 1(\text{day}) = \frac{t_6 + T_6}{1 + \frac{t_6 T_6}{\tau^2}}. \quad (9.3)$$

A straightforward calculation then shows that

$$T_6 = \frac{\tau^2}{\tau^2 - (\tau - 6)(\tau - 5)} = \frac{\tau^2}{11\tau - 30}. \quad (9.4)$$

In general one finds that

$$T_n = \frac{\tau^2}{\tau^2 - (\tau - n)(\tau - n + 1)}, \qquad (9.5)$$

or

$$T_n = \frac{\tau}{n + (n-1) - n(n-1)/\tau}. \qquad (9.6)$$

As is seen from the last formula one can neglect the last term in the denominator in the first approximation and we get the simple Eq. (9.1).

From the above one reaches the conclusion that the age of the Universe exactly equals the Hubble time in vacuum τ, i.e. 12.49 billion years, and it is a universal constant [2]. This means that the age of the Universe tomorrow will be the same as it was yesterday or today.

But this might not go along with our intuition since we usually deal with short periods of times in our daily life, and the unexperienced person will reject such a conclusion. Physics, however, deals with measurements.

9.3 Comparison with Einstein's special relativity

In fact we have a similar situation with respect to the speed of light c. When measured in vacuum, it is 300 thousands kilometers per second. If the person doing the measurement tries to decrease or increase this number by moving with a very high speed in the direction or against the direction of the propagation of light, he will find that this is impossible and he will measure the same number as before. The measurement instruments adjust themselves in such a way that the final result remains the same. In this sense the speed of light in vacuum c and the Hubble time in vacuum τ behave the same way and are both universal constants.

The similarity of the behavior of velocities of objects and those of cosmic times can also be demonstrated as follows. Suppose a rocket moves with the speed V_1 with respect to an observer on the Earth. We would like to increase that speed to V_2 as measured by the observer on the Earth. In order to achieve this, the rocket has to increase its speed not by the difference $V_2 - V_1$, but by

$$\Delta V = \frac{V_2 - V_1}{1 - \frac{V_1 V_2}{c^2}}. \tag{9.7}$$

As can easily be seen ΔV is much larger than $V_2 - V_1$ for velocities V_1 and V_2 close to that of light c. This result follows from the rule for the addition of velocities,

$$V_{1+2} = \frac{V_1 + V_2}{1 + \frac{V_1 V_2}{c^2}}, \tag{9.8}$$

a consequence of Einstein's Special Relativity Theory [3]. In cosmology, we have the analogous formula (see Eq. (2.27))

$$T_{1+2} = \frac{T_1 + T_2}{1 + \frac{T_1 T_2}{\tau^2}} \tag{9.9}$$

for the cosmic times.

9.4 References

[1] M. Carmeli, Lengths of the first days of the universe, p.628, in: *Astrophysical Ages and Time Scale*, T. von Hippel, C. Simpson and N. Manset, Editors (Proceedings of a Conference held

in Hilo, Hawaii, 5-9 February 2001) The Astronomical Society of the Pacific, Vol. 245 (2001).

[2] M. Carmeli and T. Kuzmenko, Value of the cosmological constant: theory versus experiment, in: *Proceedings of the 20th Texas Symposium on Relativistic Astrophysics*, held 10-15 December 2000, Austin, Texas, J.C. Wheeler and H. Martel, Editors (American Institute of Physics, 2001).

[3] A. Einstein, *The Meaning of Relativity*, 5th Edition (Princeton University Press, Princeton, N.J. 1955).

Appendix A

Cosmological General Relativity

The theory presented here, cosmological general relativity, uses a Riemannian four-dimensional presentation of gravitation in which the coordinates are those of Hubble, i.e. distances and velocity rather than the traditional space and time. We solve the field equations and show that there are three possibilities for the Universe to expand. The theory describes the Universe as having a three-phase evolution with a decelerating expansion, followed by a constant and an accelerating expansion, and it predicts that the Universe is now in the latter phase. It is shown, assuming $\Omega_m = 0.245$, that the time at which the Universe goes over from a decelerating to an accelerating expansion, i.e., the constant-expansion phase, occurs at 8.5 Gyr ago. Also, at that time the cosmic radiation temperature was 146K. Recent observations of distant supernovae imply, in defiance of expectations, that the Universe's growth is accelerating, contrary to what has always been assumed, that the expansion is slowing down due to gravity. Our theory confirms these recent experimental results by showing that the Universe now is definitely in a stage of accelerating expansion. The theory also predicts that now there is a positive pressure, $p = 0.034\text{g}/\text{cm}^2$, in the Universe. Although the theory has no cosmological constant, we extract its equivalence and show that $\Lambda = 1.934 \times 10^{-35}\text{s}^{-2}$. This value of Λ is in excellent agreement with the measurements obtained by the *High-Z Supernova Team* and the *Supernova Cosmology Project*. Finally it is shown that the three-dimensional space of the Universe is Euclidean, as the Boomerang experiment shows.

A.1 Preliminaries

As in classical general relativity we start our discussion in flat spacevelocity which will then be generalized to curved space.

The flat-spacevelocity cosmological metric is given by

$$ds^2 = \tau^2 dv^2 - \left(dx^2 + dy^2 + dz^2\right). \qquad (A.1)$$

Here τ is Hubble's time, the inverse of Hubble's constant, as given by measurements in the limit of zero distances and thus zero gravity. As such, τ is a constant, in fact a universal constant (its numerical value is given in Section A.8, $\tau = 12.486\text{Gyr}$). Its role in cosmology theory resembles that of c, the speed of light in vacuum, in ordinary special relativity. The velocity v is used here in the sense of cosmology, as in Hubble's law, and is usually not the time-derivative of the distance.

The Universe expansion is obtained from the metric (A.1) as a null condition, $ds = 0$. Using spherical coordinates r, θ, ϕ for the metric (A.1), and the fact that the Universe is spherically symmetric ($d\theta = d\phi = 0$), the null condition then yields $dr/dv = \tau$, or upon integration and using appropriate initial conditions, gives $r = \tau v$ or $v = H_0 r$, i.e. the Hubble law in the zero-gravity limit.

Based on the metric (A.1) a cosmological special relativity (CSR) was presented in the text (see Chapter 2). In this theory the receding velocities of galaxies and the distances between them in the Hubble expansion are united into a four-dimensional pseudo-Euclidean manifold, similarly to space and time in ordinary special relativity. The Hubble law is assumed and is written in an invariant way that enables one to derive a four-dimensional transformation which is similar to the Lorentz transformation. The parameter in the new transformation is the ratio between the cosmic time to τ (in which the cosmic time is measured backward with respect to the present time). Accordingly, the new transformation relates physical quantities at different cosmic times in the limit of weak or negligible gravitation.

The transformation between the four variables x, y, z, v and x', y', z', v' (assuming $y' = y$ and $z' = z$) is given by

$$x' = \frac{x - tv}{\sqrt{1 - t^2/\tau^2}}, \quad v' = \frac{v - tx/\tau^2}{\sqrt{1 - t^2/\tau^2}}, \quad y' = y, \ z' = z. \quad (A.2)$$

Equations (A.2) are the cosmological transformation and very much resemble the well-known Lorentz transformation. In CSR it is the relative cosmic time which takes the role of the relative velocity in Einstein's special relativity. The transformation (A.2) leaves invariant the Hubble time τ, just as the Lorentz transformation leaves invariant the speed of light in vacuum c.

A.2 Cosmology in spacevelocity

A cosmological general theory of relativity, suitable for the large-scale structure of the Universe, was subsequently developed [1-4]. In the framework of cosmological general relativity (CGR) gravitation is described by a curved four-dimensional Riemannian spacevelocity. CGR incorporates the Hubble constant τ at the outset. The Hubble law is assumed in CGR as a fundamental law. CGR, in essence, extends Hubble's law so as to incorporate gravitation in it; it is actually a *distribution theory* that relates distances and velocities between galaxies. The theory involves only measured quantities and it takes a picture of the Universe as it is at any moment. The following is a brief review of CGR as was originally given by the author in 1996 in Ref. 1.

The foundations of any gravitational theory are based on the principle of equivalence and the principle of general covariance [5]. These two principles lead immediately to the realization that gravitation should be described by a four-dimensional curved spacetime, in our theory spacevelocity, and that the field equations and the equations of motion should be written in a generally covariant form. Hence these principles were adopted in CGR also. In a four-dimensional Riemannian manifold we use a metric $g_{\mu\nu}$ and a line element $ds^2 = g_{\mu\nu}dx^\mu dx^\nu$. The difference from Einstein's general relativity is that our coordinates are: x^0 is a velocitylike coordinate (rather than a timelike coordinate), thus $x^0 = \tau v$ where τ is the Hubble time in the zero-gravity limit and v the velocity. The coordinate $x^0 = \tau v$ is the comparable to $x^0 = ct$ where c is the speed of light and t is the time in ordinary general relativity. The other three coordinates x^k, $k = 1, 2, 3$, are spacelike, just as in general relativity theory.

An immediate consequence of the above choice of coordinates is that the null condition $ds = 0$ describes the expansion of the Universe in the curved spacevelocity (generalized Hubble's law with gravitation) as compared to the propagation of light in the curved spacetime in general relativity. This means one solves the field equations (to be given in the sequel) for the metric tensor, then from the null condition $ds = 0$ one obtains immediately the dependence of the relative distances between the galaxies on their relative velocities.

As usual in gravitational theories, one equates geometry to physics. The first is expressed by means of a combination of the Ricci tensor and the Ricci scalar, and follows to be naturally either the Ricci trace-free tensor or the Einstein tensor. The Ricci trace-free tensor does not fit gravitation in general, and the Einstein tensor is a natural candidate. The physical part is expressed by the energy-momentum tensor which now has a different physical meaning from that in Einstein's theory. More important, the coupling constant that relates geometry to physics is now also different.

Accordingly the field equations are

$$G_{\mu\nu} = R_{\mu\nu} - \frac{1}{2}g_{\mu\nu}R = \kappa T_{\mu\nu}, \qquad (A.3)$$

exactly as in Einstein's theory, with κ given by $\kappa = 8\pi k/\tau^4$, (in general relativity it is given by $8\pi G/c^4$), where k is given by $k = G\tau^2/c^2$, with G being Newton's gravitational constant, and τ the Hubble constant time. When the equations of motion will be written in terms of velocity instead of time, the constant k will replace G. Using the above equations one then has $\kappa = 8\pi G/c^2\tau^2$.

The energy-momentum tensor $T^{\mu\nu}$ is constructed, along the lines of general relativity theory, with the speed of light being replaced by the Hubble constant time. If ρ is the average mass density of the Universe, then it will be assumed that $T^{\mu\nu} = \rho u^\mu u^\nu$, where $u^\mu = dx^\mu/ds$ is the four-velocity. In general relativity theory one takes $T_0^0 = \rho$. In Newtonian gravity one has the Poisson equation $\nabla^2 \phi = 4\pi G\rho$. At points where $\rho = 0$ one solves the vacuum Einstein field equations in general relativity and the Laplace equation $\nabla^2 \phi = 0$ in Newtonian gravity. In both theories a null (zero) solution is allowed as a trivial case. In cosmology, however, there exists no situation at which ρ can be zero because the Universe is filled with matter. In order to be able to have zero on the right-hand side of Eq. (A.3) one takes T_0^0 not as equal to ρ, but to $\rho_{eff} = \rho - \rho_c$, where ρ_c is the critical mass density, a *constant* in CGR given by $\rho_c = 3/8\pi G\tau^2$, whose value is $\rho_c \approx 10^{-29}\text{g/cm}^3$, a few hydrogen atoms per cubic meter. Accordingly one takes

$$T^{\mu\nu} = \rho_{eff} u^\mu u^\nu; \quad \rho_{eff} = \rho - \rho_c \qquad (A.4)$$

for the energy-momentum tensor.

In the next sections we apply CGR to obtain the accelerating expanding Universe and related subjects.

A.3 Gravitational field equations

In the four-dimensional spacevelocity the spherically symmetric metric is given by

$$ds^2 = \tau^2 dv^2 - e^\mu dr^2 - R^2 \left(d\theta^2 + \sin^2\theta d\phi^2\right), \qquad (A.5)$$

where μ and R are functions of v and r alone, and comoving coordinates $x^\mu = (x^0, x^1, x^2, x^3) = (\tau v, r, \theta, \phi)$ have been used. With the above choice of coordinates, the zero-component of the geodesic equation becomes an identity, and since r, θ and ϕ are constants along the geodesics, one has $dx^0 = ds$ and therefore

$$u^\alpha = u_\alpha = (1, 0, 0, 0). \qquad (A.6)$$

The metric (A.5) shows that the area of the sphere $r = constant$ is given by $4\pi R^2$ and that R should satisfy $R' = \partial R/\partial r > 0$. The possibility that $R' = 0$ at a point r_0 is excluded since it would allow the lines $r = constants$ at the neighboring points r_0 and $r_0 + dr$ to coincide at r_0, thus creating a caustic surface at which the comoving coordinates break down.

As has been shown in the previous sections the Universe expands by the null condition $ds = 0$, and if the expansion is spherically symmetric one has $d\theta = d\phi = 0$. The metric (A.5) then yields

$$\tau^2 dv^2 - e^\mu dr^2 = 0, \qquad (A.7)$$

Gravitational field equations

thus

$$\frac{dr}{dv} = \tau e^{-\mu/2}. \tag{A.8}$$

This is the differential equation that determines the Universe expansion. In the following we solve the gravitational field equations in order to find out the function $\mu(r.v)$.

The gravitational field equations (A.3), written in the form

$$R_{\mu\nu} = \kappa \left(T_{\mu\nu} - \frac{1}{2} g_{\mu\nu} T \right), \tag{A.9}$$

where

$$T_{\mu\nu} = \rho_{eff} u_\mu u_\nu + p \left(u_\mu u_\nu - g_{\mu\nu} \right), \tag{A.10}$$

with $\rho_{eff} = \rho - \rho_c$ and $T = T_{\mu\nu} g^{\mu\nu}$, are now solved. Using Eq. (A.6) one finds that the only nonvanishing components of $T_{\mu\nu}$ are $T_{00} = \tau^2 \rho_{eff}$, $T_{11} = c^{-1} \tau p e^\mu$, $T_{22} = c^{-1} \tau p R^2$ and $T_{33} = c^{-1} \tau p R^2 \sin^2 \theta$, and that $T = \tau^2 \rho_{eff} - 3c^{-1} \tau p$.

The only nonvanishing components of the Ricci tensor yield (dots and primes denote differentiation with respect to v and r, respectively), using Eq. (A.9), the following field equations:

$$R_{00} = -\frac{1}{2}\ddot{\mu} - \frac{2}{R}\ddot{R} - \frac{1}{4}\dot{\mu}^2 = \frac{\kappa}{2}\left(\tau^2 \rho_{eff} + 3c^{-1}\tau p\right), \tag{A.11a}$$

$$R_{01} = \frac{1}{R}R'\dot{\mu} - \frac{2}{R}\dot{R}' = 0, \tag{A.11b}$$

$$R_{11} = e^\mu \left(\frac{1}{2}\ddot{\mu} + \frac{1}{4}\dot{\mu}^2 + \frac{1}{R}\dot{\mu}\dot{R} \right) + \frac{1}{R}\left(\mu'R' - 2R'' \right)$$

$$= \frac{\kappa}{2} e^\mu \left(\tau^2 \rho_{eff} - c^{-1}\tau p \right), \tag{A.11c}$$

$$R_{22} = R\ddot{R} + \frac{1}{2}R\dot{R}\dot{\mu} + \dot{R}^2 + 1 - e^{-\mu}\left(RR'' - \frac{1}{2}RR'\mu' + R'^2 \right)$$

$$= \frac{\kappa}{2} R^2 \left(\tau^2 \rho_{eff} - c^{-1}\tau p \right), \tag{A.11d}$$

$$R_{33} = \sin^2\theta R_{22} = \frac{\kappa}{2} R^2 \sin^2\theta \left(\tau^2 \rho_{eff} - c^{-1}\tau p\right). \quad (A.11e)$$

The field equations obtained for the components 00, 01, 11, and 22 (the 33 component contributes no new information) are given by

$$-\ddot{\mu} - \frac{4}{R}\ddot{R} - \frac{1}{2}\dot{\mu}^2 = \kappa\left(\tau^2 \rho_{eff} + 3c^{-1}\tau p\right), \quad (A.12)$$

$$2\dot{R}' - R'\dot{\mu} = 0, \quad (A.13)$$

$$\ddot{\mu} + \frac{1}{2}\dot{\mu}^2 + \frac{2}{R}\dot{R}\dot{\mu} + e^{-\mu}\left(\frac{2}{R}R'\mu' - \frac{4}{R}R''\right) = \kappa\left(\tau^2 \rho_{eff} - c^{-1}\tau p\right) \quad (A.14)$$

$$\frac{2}{R}\ddot{R} + 2\left(\frac{\dot{R}}{R}\right)^2 + \frac{1}{R}\dot{R}\dot{\mu} + \frac{2}{R^2} + e^{-\mu}\left[\frac{1}{R}R'\mu' - 2\left(\frac{R'}{R}\right)^2 - \frac{2}{R}R''\right]$$

$$= \kappa\left(\tau^2 \rho_{eff} - c^{-1}\tau p\right). \quad (A.15)$$

It is convenient to eliminate the term with the second velocity-derivative of μ from the above equations. This can easily be done, and combinations of Eqs. (A.12)–(A.15) then give the following set of three independent field equations:

$$e^{\mu}\left(2R\ddot{R} + \dot{R}^2 + 1\right) - R'^2 = -\kappa\tau c^{-1} e^{\mu} R^2 p, \quad (A.16)$$

$$2\dot{R}' - R'\dot{\mu} = 0, \quad (A.17)$$

$$e^{-\mu}\left[\frac{1}{R}R'\mu' - \left(\frac{R'}{R}\right)^2 - \frac{2}{R}R''\right] + \frac{1}{R}\dot{R}\dot{\mu} + \left(\frac{\dot{R}}{R}\right)^2 + \frac{1}{R^2}$$

$$= \kappa\tau^2 \rho_{eff}, \quad (A.18)$$

other equations being trivial combinations of (A.16)–(A.18).

A.4 Solution of the field equations

The solution of Eq. (A.17) satisfying the condition $R' > 0$ is given by

$$e^\mu = \frac{R'^2}{1 + f(r)}, \qquad (A.19)$$

where $f(r)$ is an arbitrary function of the coordinate r and satisfies the condition $f(r) + 1 > 0$. Substituting (A.19) in the other two field equations (A.16) and (A.18) then gives

$$2R\ddot{R} + \dot{R}^2 - f = -\kappa c^{-1} \tau R^2 p, \qquad (A.20)$$

$$\frac{1}{RR'}\left(2\dot{R}\dot{R}' - f'\right) + \frac{1}{R^2}\left(\dot{R}^2 - f\right) = \kappa \tau^2 \rho_{eff}, \qquad (A.21)$$

respectively.

The simplest solution of the above two equations, which satisfies the condition $R' = 1 > 0$, is given by

$$R = r. \qquad (A.22)$$

Using Eq. (A.22) in Eqs. (A.20) and (A.21) gives

$$f(r) = \kappa c^{-1} \tau p r^2, \qquad (A.23)$$

and

$$f' + \frac{f}{r} = -\kappa \tau^2 \rho_{eff} r, \qquad (A.24)$$

respectively. The solution of Eq. (A.24) is the sum of the solutions of the homogeneous equation

$$f' + \frac{f}{r} = 0, \qquad (A.25)$$

and a particular solution of Eq. (A.24). These are given by

$$f_1 = -\frac{2Gm}{c^2 r}, \qquad (A.26)$$

and
$$f_2 = -\frac{\kappa}{3}\tau^2 \rho_{eff} r^2. \tag{A.27}$$

The solution f_1 represents a particle at the origin of coordinates and as such is not relevant to our problem. We take, accordingly, f_2 as the general solution,

$$f(r) = -\frac{\kappa}{3}\tau^2 \rho_{eff} r^2 = -\frac{\kappa}{3}\tau^2 (\rho - \rho_c) r^2$$

$$= -\frac{\kappa}{3}\tau^2 \rho_c \left(\frac{\rho}{\rho_c} - 1\right) r^2. \tag{A.28}$$

Using the values of $\kappa = 8\pi G/c^2\tau^2$ and $\rho_c = 3/8\pi G\tau^2$, we obtain

$$f(r) = \frac{1-\Omega}{c^2\tau^2} r^2, \tag{A.29}$$

where $\Omega = \rho/\rho_c$.

The two solutions given by Eqs. (A.23) and (A.29) for $f(r)$ can now be equated, giving

$$p = \frac{1-\Omega}{\kappa c\tau^3} = \frac{c}{\tau}\frac{1-\Omega}{8\pi G} = 4.544(1-\Omega) \times 10^{-2} \text{g/cm}^2. \tag{A.30}$$

Furthermore, from Eqs. (A.19) and (A.22) we find that

$$e^{-\mu} = 1 + f(r) = 1 + \tau c^{-1}\kappa p r^2 = 1 + \frac{1-\Omega}{c^2\tau^2} r^2. \tag{A.31}$$

It will be recalled that the Universe expansion is determined by Eq. (A.8), $dr/dv = \tau e^{-\mu/2}$. The only thing that is left to be determined is the signs of $(1-\Omega)$ or the pressure p.

Thus we have

$$\frac{dr}{dv} = \tau\sqrt{1 + \kappa\tau c^{-1}pr^2} = \tau\sqrt{1 + \frac{1-\Omega}{c^2\tau^2}r^2}. \tag{A.32}$$

For simplicity we confine ourselves to the linear approximation, thus Eq. (A.32) yields

$$\frac{dr}{dv} = \tau\left(1 + \frac{\kappa}{2}\tau c^{-1}pr^2\right) = \tau\left[1 + \frac{1-\Omega}{2c^2\tau^2}r^2\right]. \tag{A.33}$$

A.5 Classification of universes

The second term in the square bracket in the above equation represents the deviation due to gravity from the standard Hubble law. For without that term, Eq. (A.33) reduces to $dr/dv = \tau$, thus $r = \tau v + const$. The constant can be taken zero if one assumes, as usual, that at $r = 0$ the velocity should also vanish. Thus $r = \tau v$, or $v = H_0 r$ (since $H_0 \approx 1/\tau$). Accordingly, the equation of motion (A.33) describes the expansion of the Universe when $\Omega = 1$, namely when $\rho = \rho_c$. The equation then coincides with the standard Hubble law.

The equation of motion (A.33) can easily be integrated exactly by the substitutions

$$\sin\chi = \sqrt{\frac{(\Omega-1)}{2}\frac{r}{2c\tau}}; \quad \Omega > 1, \qquad (A.34a)$$

$$\sinh\chi = \sqrt{\frac{(1-\Omega)}{2}\frac{r}{2c\tau}}; \quad \Omega < 1. \qquad (A.34b)$$

One then obtains, using Eqs. (A.33) and (A.34),

$$dv = cd\chi/(\Omega-1)^{1/2}\cos\chi; \quad \Omega > 1, \qquad (A.35a)$$

$$dv = cd\chi/(1-\Omega)^{1/2}\cosh\chi; \quad \Omega < 1. \qquad (A.35b)$$

We give below the exact solutions for the expansion of the Universe for each of the cases, $\Omega > 1$ and $\Omega < 1$. As will be seen, the case of $\Omega = 1$ can be obtained at the limit $\Omega \to 1$ from both cases.

The case $\Omega > 1$. From Eq. (A.35a) we have

$$\int dv = \frac{c}{\sqrt{(\Omega-1)/2}}\int\frac{d\chi}{\cos\chi}, \qquad (A.36)$$

where $\sin\chi = r/a$, and $a = c\tau\sqrt{(\Omega-1)/2}$. A simple calculation gives [6]

$$\int \frac{d\chi}{\cos\chi} = \ln\left|\frac{1+\sin\chi}{\cos\chi}\right|. \quad (A.37)$$

A straightforward calculation then gives

$$v = \frac{a}{2\tau}\ln\left|\frac{1+r/a}{1-r/a}\right|. \quad (A.38)$$

As is seen, when $r \to 0$ then $v \to 0$ and using the L'Hospital lemma, $v \to r/\tau$ as $a \to 0$ (and thus $\Omega \to 1$).

The case $\Omega < 1$. From Eq. (A.35b) we now have

$$\int dv = \frac{c}{\sqrt{(1-\Omega)/2}}\int \frac{d\chi}{\cosh\chi}, \quad (A.39)$$

where $\sinh\chi = r/b$ and $b = c\tau\sqrt{(1-\Omega)/2}$. A straightforward calculation then gives [6]

$$\int \frac{d\chi}{\cosh\chi} = \arctan e^\chi. \quad (A.40)$$

We then obtain

$$\cosh\chi = \sqrt{1+\frac{r^2}{b^2}}, \quad (A.41)$$

$$e^\chi = \sinh\chi + \cosh\chi = \frac{r}{b} + \sqrt{1+\frac{r^2}{b^2}}. \quad (A.42)$$

Equations (A.39) and (A.40) now give

$$v = \frac{2c}{\sqrt{(1-\Omega)/2}}\arctan e^\chi + K, \quad (A.43)$$

where K is an integration constant which is determined by the requirement that at $r = 0$, v should then be zero. We obtain

$$K = -\pi c/2\sqrt{(1-\Omega)/2}, \quad (A.44)$$

and thus
$$v = \frac{2c}{\sqrt{(1-\Omega)/2}}\left(\arctan e^{\chi} - \frac{\pi}{4}\right). \quad (A.45)$$

A straightforward calculation then gives
$$v = \frac{b}{\tau}\left\{2\arctan\left(\frac{r}{b} + \sqrt{1 + \frac{r^2}{b^2}}\right) - \frac{\pi}{2}\right\}. \quad (A.46)$$

As for the case $\Omega > 1$ one finds that $v \to 0$ when $r \to 0$, and again, using L'Hospital lemma, $r = \tau v$ when $b \to 0$ (and thus $\Omega \to 1$).

A.6 Physical meaning

To see the physical meaning of these solutions, however, one does not need the exact solutions. Rather, it is enough to write down the solutions in the lowest approximation in τ^{-1}. One obtains, by differentiating Eq. (A.33) with respect to v, for $\Omega > 1$,
$$d^2r/dv^2 = -kr; \quad k = \frac{(\Omega - 1)}{2c^2}, \quad (A.47)$$

the solution of which is
$$r(v) = A \sin\alpha\frac{v}{c} + B\cos\alpha\frac{v}{c}, \quad (A.48)$$

where $\alpha^2 = (\Omega - 1)/2$ and A and B are constants. The latter can be determined by the initial condition $r(0) = 0 = B$ and $dr(0)/dv = \tau = A\alpha/c$, thus
$$r(v) = \frac{c\tau}{\alpha}\sin\alpha\frac{v}{c}. \quad (A.49)$$

This is obviously a closed Universe, and presents a decelerating expansion.

For $\Omega < 1$ we have

$$d^2r/dv^2 = \frac{(1-\Omega)\,r}{2c^2}, \qquad (A.50)$$

whose solution, using the same initial conditions, is

$$r(v) = \frac{c\tau}{\beta}\sinh\beta\frac{v}{c}, \qquad (A.51)$$

where $\beta^2 = (1-\Omega)/2$. This is now an open accelerating Universe. For $\Omega = 1$ we have, of course, $r = \tau v$.

A.7 The accelerating universe

We finally determine which of the three cases of expansion is the one at present epoch of time. To this end we have to write the solutions (A.49) and (A.51) in ordinary Hubble's law form $v = H_0 r$. Expanding Eqs. (A.49) and (A.51) into power series in v/c and keeping terms up to the second order, we obtain

$$r = \tau v\left(1 - \alpha^2 v^2/6c^2\right), \qquad (A.52a)$$

$$r = \tau v\left(1 + \beta^2 v^2/6c^2\right), \qquad (A.52b)$$

for $\Omega > 1$ and $\Omega < 1$, respectively. Using now the expressions for α and β, Eqs. (A.52) then reduce into the single equation

$$r = \tau v\left[1 + (1-\Omega)\,v^2/6c^2\right]. \qquad (A.53)$$

Inverting now this equation by writing it as $v = H_0 r$, we obtain in the lowest approximation

$$H_0 = h\left[1 - (1-\Omega)\,v^2/6c^2\right], \qquad (A.54)$$

where $h = \tau^{-1}$. To the same approximation one also obtains

$$H_0 = h\left[1 - (1-\Omega)\,z^2/6\right] = h\left[1 - (1-\Omega)\,r^2/6c^2\tau^2\right], \qquad (A.55)$$

The accelerating universe

Table A.1: The Cosmic Times with respect to the Big Bang, the Cosmic Temperature and the Cosmic Pressure for each of the Curves in Fig. A.1.

Curve No*.	Ω_m	Time (τ)	Time (Gyr)	Temp. (K)	Pressure (g/cm^2)
		DECELERATING EXPANSION			
1	100	3.1×10^{-6}	3.87×10^{-5}	1096	-4.499
2	25	9.8×10^{-5}	1.22×10^{-3}	195.0	-1.091
3	10	3.0×10^{-4}	3.75×10^{-3}	111.5	-0.409
4	5	1.2×10^{-3}	1.50×10^{-2}	58.20	-0.182
5	1.5	1.3×10^{-2}	1.62×10^{-1}	16.43	-0.023
		CONSTANT EXPANSION			
6	1	3.0×10^{-2}	3.75×10^{-1}	11.15	0
		ACCELERATING EXPANSION			
7	0.5	1.3×10^{-1}	1.62	5.538	$+0.023$
8	0.245	1.0	12.50	2.730	$+0.034$

*The calculations are made using Carmeli's cosmological transformation, Eq. (A.2), that relates physical quantities at different cosmic times when gravity is extremely weak.

For example, we denote the temperature by θ, and the temperature at the present time by θ_0, we then have

$$\theta = \frac{\theta_0}{\sqrt{1 - \frac{t^2}{\tau^2}}} = \frac{\theta_0}{\sqrt{1 - \frac{(\tau - T)^2}{\tau^2}}} = \frac{2.73K}{\sqrt{\frac{2\tau T - T^2}{\tau^2}}} = \frac{2.73K}{\sqrt{\frac{T}{\tau}\left(2 - \frac{T}{\tau}\right)}},$$

where T is the time with respect to B.B.

The formula for the pressure is given by Eq. (A.30), $p = c(1-\Omega)/8\pi G\tau$. Using $c = 3 \times 10^{10}$cm/s, $\tau = 3.938 \times 10^{17}$s and $G = 6.67 \times 10^{-8}$cm^3/gs^2, we obtain

$$p = 4.544 \times 10^{-2} (1 - \Omega) \, \text{g/cm}^2.$$

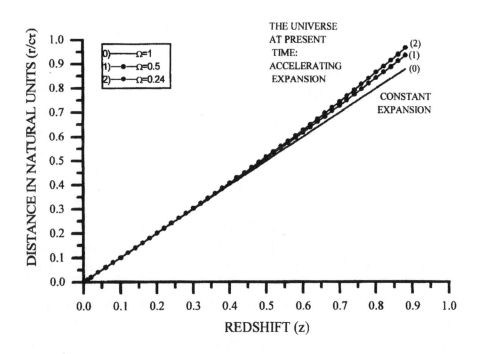

Figure A.2 Hubble's diagram of the Universe at the present phase of evolution with accelerating expansion. (Source: Behar & Carmeli, Ref. 2)

The accelerating universe 133

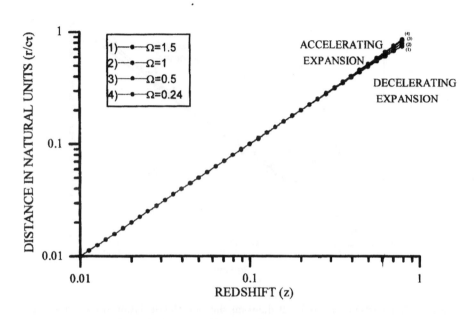

Figure A.3 Hubble's diagram describing decelerating, constant and accelerating expansions in a logarithmic scale. (Source: Behar & Carmeli, Ref. 2)

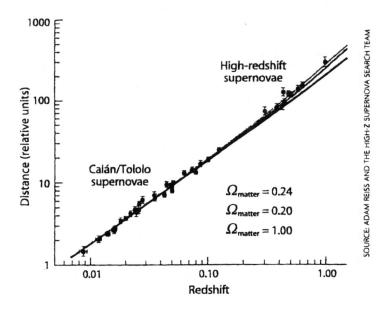

Figure A.4 Distance vs. redshift diagram showing the deviation from a constant toward an accelerating expansion. (Source: Riess *et al.*, Ref. 11)

The accelerating universe

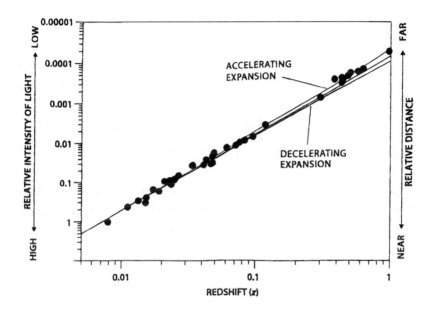

Figure A.5 Relative intensity of light and relative distance vs. redshift. (Source: Riess *et al.*, Ref. 11)

Figures A.2 and A.3 (Behar & Carmeli) show the Hubble diagrams for the distance-redshift relationship predicted by the theory for the accelerating expanding Universe at the present time, and Figures A.4 and A.5 (Riess *et al.*) show the experimental results.

Our estimate for h, based on published data, is $h \approx 80$ km/sec-Mpc. Assuming $\tau^{-1} \approx 80$ km/sec-Mpc, Eq. (A.55) then gives

$$H_0 = h \left[1 - 1.3 \times 10^{-4} \left(1 - \Omega\right) r^2\right], \quad (A.56)$$

where r is in Mpc. A computer best-fit can then fix both h and Ω_m.

To summarize, a theory of cosmology has been presented in which the dynamical variables are those of Hubble, i.e. distances and velocities. The theory describes the Universe as having a three-phase evolution with a decelerating expansion, followed by a constant and an accelerating expansion, and it predicts that the Universe is now in the latter phase. As the density of matter decreases, while the Universe is at the decelerating phase, it does not have enough time to close up to a big crunch. Rather, it goes to the constant-expansion phase, and then to the accelerating stage. As we have seen, the equation obtained for the Universe expansion, Eq. (A.51), is very simple.

A.8 Theory versus experiment

The Einstein gravitational field equations with the added cosmological term are [8]:

$$R_{\mu\nu} - \frac{1}{2} g_{\mu\nu} R + \Lambda g_{\mu\nu} = \kappa T_{\mu\nu}, \quad (A.57)$$

where Λ is the cosmological constant, the value of which is supposed to be determined by experiment. In Eq. (A.57) $R_{\mu\nu}$ and

Theory versus experiment

R are the Ricci tensor and scalar, respectively, $\kappa = 8\pi G$, where G is Newton's constant and the speed of light is taken as unity.

Recently the two groups (the *Supernovae Cosmology Project* and the *High-Z Supernova Team*) concluded that the expansion of the Universe is accelerating [9-15]. The two groups had discovered and measured moderately high redshift ($0.3 < z < 0.9$) supernovae, and found that they were fainter than what one would expect them to be if the cosmos expansion were slowing down or constant. Both teams obtained

$$\Omega_m \approx 0.3, \quad \Omega_\Lambda \approx 0.7, \quad (A.58)$$

and ruled out the traditional (Ω_m, Ω_Λ)=(1, 0) Universe. Their value of the density parameter Ω_Λ corresponds to a cosmological constant that is small but, nevertheless, nonzero and positive,

$$\Lambda \approx 10^{-52} \text{m}^{-2} \approx 10^{-35} \text{s}^{-2}. \quad (A.59)$$

In previous sections a four-dimensional cosmological theory (CGR) was presented. Although the theory has no cosmological constant, it predicts that the Universe accelerates and hence it has the equivalence of a positive cosmological constant in Einstein's general relativity. In the framework of this theory (see Section A.2) the zero-zero component of the field equations (A.3) is written as

$$R_0^0 - \frac{1}{2}\delta_0^0 R = \kappa \rho_{eff} = \kappa(\rho - \rho_c), \quad (A.60)$$

where $\rho_c = 3/\kappa\tau^2$ is the critical mass density and τ is Hubble's time in the zero-gravity limit.

Comparing Eq. (A.60) with the zero-zero component of Eq. (A.57), one obtains the expression for the cosmological constant of general relativity,

$$\Lambda = \kappa\rho_c = 3/\tau^2. \quad (A.61)$$

To find out the numerical value of τ we use the relationship between $h = \tau^{-1}$ and H_0 given by Eq. (A.55) (CR denote values according to Cosmological Relativity):

$$H_0 = h\left[1 - \left(1 - \Omega_m^{CR}\right) z^2/6\right], \qquad (A.62)$$

where $z = v/c$ is the redshift and $\Omega_m^{CR} = \rho_m/\rho_c$ with $\rho_c = 3h^2/8\pi G$. (Notice that our $\rho_c = 1.194 \times 10^{-29} \text{g/cm}^3$ is different from the standard ρ_c defined with H_0.) The redshift parameter z determines the distance at which H_0 is measured. We choose $z = 1$ and take for

$$\Omega_m^{CR} = 0.245, \qquad (A.63)$$

its value at the present time (see Table A.1) (corresponds to 0.32 in the standard theory), Eq. (A.62) then gives

$$H_0 = 0.874h. \qquad (A.64)$$

At the value $z = 1$ the corresponding Hubble parameter H_0 according to the latest results from HST can be taken [16] as $H_0 = 70\text{km/s-Mpc}$, thus $h = (70/0.874)\text{km/s-Mpc}$, or

$$h = 80.092\text{km/s-Mpc}, \qquad (A.65)$$

and

$$\tau = 12.486\text{Gyr} = 3.938 \times 10^{17}\text{s}. \qquad (A.66)$$

What is left is to find the value of Ω_Λ^{CR}. We have $\Omega_\Lambda^{CR} = \rho_c^{ST}/\rho_c$, where $\rho_c^{ST} = 3H_0^2/8\pi G$ and $\rho_c = 3h^2/8\pi G$. Thus $\Omega_\Lambda^{CR} = (H_0/h)^2 = 0.874^2$, or

$$\Omega_\Lambda^{CR} = 0.764. \qquad (A.67)$$

As is seen from Eqs. (A.63) and (A.67) one has

$$\Omega_T = \Omega_m^{CR} + \Omega_\Lambda^{CR} = 0.245 + 0.764 = 1.009 \approx 1, \qquad (A.68)$$

which means the Universe is Euclidean.

As a final result we calculate the cosmological constant according to Eq. (A.61). One obtains

$$\Lambda = 3/\tau^2 = 1.934 \times 10^{-35} \text{s}^{-2}. \qquad (A.69)$$

Our results confirm those of the supernovae experiments and indicate on the existence of the dark energy as has recently received confirmation from the Boomerang cosmic microwave background experiment [17,18], which showed that the Universe is Euclidean.

A.9 Concluding remarks

In this Appendix the cosmological general relativity, a relativistic theory in spacevelocity, has been presented and applied to the problem of the expansion of the Universe. The theory, which predicts a positive pressure for the Universe now, describes the Universe as having a three-phase evolution: decelerating, constant and accelerating expansion, but it is now in the latter stage. Furthermore, the cosmological constant that was extracted from the theory agrees with the experimental result. Finally, it has also been shown that the three-dimensional spatial space of the Universe is Euclidean, again in agreement with observations.

The most direct evidence that the Universe expansion is accelerating and propelled by "dark energy", is provided by the faintness of Type Ia supernovae (SNe Ia) at $z \approx 0.5$ [10,14]. Beyond the redshift range of $0.5 < z < 1$, the Universe was more compact and the attraction of matter was stronger than the repulsive dark energy. At $z > 1$ the expansion of the Universe should have been decelerating [19]. At $z \geq 1$ one would expect an apparent brightness increase of SNe Ia relative to what is supposed to be for a non-decelerating Universe [20].

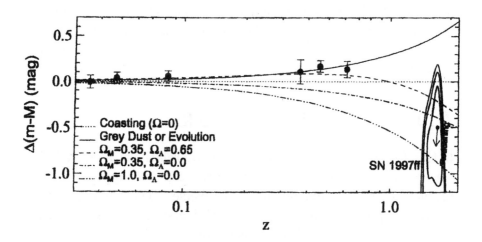

Figure A.6 Hubble diagram of SNe Ia minus an empty (i.e., "empty" $\Omega = 0$) universe compared to cosmological and astrophysical models. The measurements of SN 1997ff are inconsistent with astrophysical effects which could mimic previous evidence for an accelerating Universe from SNe Ia at $z \approx 0.5$. (Source: Riess *et al.*, Ref. 20)

Concluding remarks

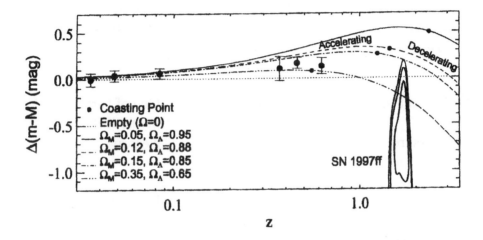

Figure A.7 Same as Figure A.6 with the inclusion of a family of plausible, flat Ω_Λ cosmologies. The transition redshift (i.e., the coasting point) between the accelerating and decelerating phases is indicated and is given as $[2\Omega_\Lambda/\Omega_M]^{1/3} - 1$. SN 1997ff is seen to lie within the epoch of deceleration. This conclusion is drawn from the result that the apparent brightness of SN 1997ff is inconsistent with values of $\Omega_\Lambda \geq 0.9$ and hence a transition redshift greater than that of SN 1997ff. (Source: Riess *et al.*, Ref. 20)

Table A.2: Cosmological parameters in cosmological general relativity and in standard theory

	COSMOLOGICAL RELATIVITY	STANDARD THEORY
Theory type	Spacevelocity	Spacetime
Expansion type	Tri-phase: decelerating, constant, accelerating	One phase
Present expansion	Accelerating (predicted)	One of three possibilities
Pressure	0.034g/cm^2	Negative
Cosmological constant	1.934×10^{-35}s^{-2} (predicted)	Depends
$\Omega_T = \Omega_m + \Omega_\Lambda$	1.009	Depends
Constant-expansion occurs at	8.5Gyr ago*	No prediction
Constant-expansion duration	Fraction of second	Not known
Temperature at constant expansion	146K*	No prediction

*Adjusted to include gravity (see Appendix C)

Recently [19,20], more confirmation to the Universe accelerating expansion came from the most distant supernova, SN 1997ff, that was recorded by the Hubble Space Telescope. As has been pointed out before, if we look back far enough, we should find a decelerating expansion (curves 1-5 in Figure A.1). Beyond $z = 1$

one should see an earlier time when the mass density was dominant. The measurements obtained from SN 1997ff's redshift and brightness provide a direct proof for the transition from past decelerating to present accelerating expansion (see Figures A.6 and A.7, Riess *et al.*). The measurements also exclude the possibility that the acceleration of the Universe is not real but is due to other astrophysical effects such as dust.

Table A.2 gives some of the cosmological parameters obtained here and in the standard theory.

A.10 References

1. M. Carmeli, *Commun. Theor. Phys.* **5**, 159 (1996).
2. S. Behar and M. Carmeli, *Intern. J. Theor. Phys.* **39**, 1375 (2000). (astro-ph/0008352)
3. M. Carmeli and S. Behar, Cosmological general relativity, pp. 5–26, in: *Quest for Mathematical Physics*, T.M. Karade, *et al.* Editors, New Delhi (2000).
4. M. Carmeli and S. Behar, Cosmological relativity: a general relativistic theory for the accelerating universe, Talk given at Dark Matter 2000, Los Angeles, February 2000, pp.182–191, in: *Sources and Detection of Dark Matter/Energy in the Universe*, D. Cline, Ed., Springer (2001).
5. A. Einstein, *The Meaning of Relativity*, 5th Edition (Princeton Univ. Press, Princeton, 1955).
6. I.S. Gradshteyn and I.M. Ryshik, *Table of Integrals, Series and Products* (Academic Press, New York; 1980).
7. P.J.E. Peebles, Status of the big bang cosmology, in: *Texas/Pascos 92: Relativistic Astrophysics and Particle Cosmology*, C.W. Akerlof and M.A. Srednicki, Editors (New York Academy of Sciences, New York, 1993), p. 84.
8. M. Carmeli, *Classical Fields: General Relativity and Gauge Theory* (Wiley, New York, 1982).

9. P.M. Garnavich et al., *Astrophys. J.* **493**, L53 (1998). [Hi-Z Supernova Team Collaboration (astro-ph/9710123)].
10. B.P. Schmidt et al., *Astrophys. J.* **507**, 46 (1998). [Hi-Z Supernova Team Collaboration (astro-ph/9805200)].
11. A.G. Riess et al., *Astronom. J.* **116**, 1009 (1998). [Hi-Z Supernova Team Collaboration (astro-ph/9805201)].
12. P.M. Garnavich et al., *Astrophys. J.* **509**, 74 (1998). [Hi-Z Supernova Team Collaboration (astro-ph/9806396)].
13. S. Perlmutter et al., *Astrophys. J.* **483**, 565 (1997). [Supernova Cosmology Project Collaboration (astro-ph/9608192)].
14. S. Perlmutter et al., *Nature* **391**, 51 (1998). [Supernova Cosmology Project Collaboration (astro-ph/9712212)].
15. S. Perlmutter et al., *Astrophys. J.* **517**, 565 (1999). [Supernova Cosmology Project Collaboration (astro-ph/9812133)].
16. W.L. Freedman et al., Final results from the Hubble Space Telescope, Talk given at the 20th Texas Symposium on Relativistic Astrophysics, Austin, Texas 10-15 December 2000. (astro-ph/0012376)
17. P. de Bernardis et al., *Nature* **404**, 955 (2000). (astro-ph/0004404)
18. J.R. Bond et al., in *Proc. IAU Symposium* 201 (2000). (astro-ph/0011378)
19. A.V. Filippenko and A.G. Riess, p.227 in: *Particle Physics and Cosmology: Second Tropical Workshop*, J.F. Nieves, Editor (AIP, New York, 2000).
20. A.G. Riess et al., *Astrophys. J.*, in press. (astro-ph/0104455)

Appendix B

Five-Dimensional Brane World Theory

A five-dimensional cosmological theory of gravitation that unifies space, time and velocity is presented. It is actually a unification of Einstein's general relativity and the author's cosmological general relativity given in Appendix A. Within the framework of this theory we first discuss some general aspects of the Universe in five dimensions. We then find the equations of motion of the expanding Universe and show that it is accelerating. This followed by dealing with the important problem of halo dark matter around galaxies by deriving the equations of motion of a star moving around the field of a spherically-symmetric galaxy, and the Tully-Fisher formula is obtained. The cosmological constant is subsequently discussed: our theory predicts that $\Lambda = 1.934 \times 10^{-35} \text{s}^{-2}$, in agreement with experimental results. Finally we derive a formula for the cosmological redshift in which appears the expression $(1 - \Omega_M)$, thus enabling us to determine the kind of the Universe by means of the cosmological redshift. We find that Ω_M should be less than 1 in order not to contradict redshift measurements, and therefore the Universe is open and its spatial part is Euclidean.

B.1 Introduction

In this appendix we present a five-dimensional cosmological theory of space, time and velocity. The added extra dimension of velocity to the usual four-dimensional spacetime will be evident in the sequel. Before introducing the theory we have to deal, as usual, with coordinate systems in cosmology. Other important basic issues will be dealt later on.

B.1.1 Cosmic coordinate systems: The Hubble transformation

We will use *cosmic coordinate systems* that fill up spacetime. Given one system x, there is another one x' that differs from the original one by a *Hubble transformation*

$$x' = x + t_1 v, \quad t_1 = \text{constant}, \qquad (B.1)$$

where v is a velocity parameter, and y and z are kept unchanged. A third system will be given by another Hubble transformation,

$$x'' = x' + t_2 v = x + (t_1 + t_2) v. \qquad (B.2)$$

The cosmic coordinate systems are similar to the inertial coordinate systems, but now the velocity parameter takes over the time parameter and vice versa. The analogous Galileo transformation to Eq. (B.2) that relates inertial coordinate systems is

Introduction

given, as is known, by

$$x'' = x' + v_2 t = x + (v_1 + v_2)t. \qquad (B.3)$$

The Universe expansion is also given by a formula of the above kind:

$$x' = x + \tau v, \qquad (B.4)$$

where $\tau = H_0^{-1}$ in the limit of zero distance, and thus a universal constant. (Its value is calculated in Subsection B.5.4 as $\tau = 12.486\text{Gyr}$.) However, the Universe expansion is apparently incompatible with the Hubble spacetime transformation, namely one cannot add them. Thus, if we have

$$x'' = x' + tv, \quad x' = \tau v, \qquad (B.5)$$

then

$$x'' \neq (\tau + t) v. \qquad (B.6)$$

Rather, it is always

$$x'' = \tau v. \qquad (B.7)$$

The above can be looked upon as a postulate of the theory.

This situation is like the one we have with the propagation of light,

$$x'' \neq (c + v) t, \qquad (B.8)$$

but it is always

$$x'' = ct \qquad (B.9)$$

in all inertial coordinate systems, and where c is the speed of light in vacuum.

The constancy of the speed of light and the validity of the laws of nature in inertial coordinate systems, though they are both experimentally valid, they are apparently not compatible with each other. We have the same situation in cosmology; the constancy of the Hubble constant in the zero-distance limit, and the validity of the laws of nature in cosmic coordinate systems, though both are valid, they are apparently incompatible with each other.

B.1.2 Lorentz-like cosmological transformation

In the case of light propagation, one has to abandon the Galileo transformation in favor of the Lorentz transformation. In cosmology one has to give up the Hubble transformation (B.1) in favor of a new Lorentz-like cosmological transformation given by [1]

$$x' = \frac{x - tv}{\sqrt{1 - t^2/\tau^2}}, \quad v' = \frac{v - tx/\tau^2}{\sqrt{1 - t^2/\tau^2}}, \quad y' = y, \quad z' = z, \tag{B.10}$$

for the case with fixed y and z.

As is well known, the flat spacetime line element in special relativity is given by

$$ds^2 = c^2 dt^2 - (dx^2 + dy^2 + dz^2). \tag{B.11}$$

The cosmological flat spacevelocity line element is given, accordingly, by

$$ds^2 = \tau^2 dv^2 - (dx^2 + dy^2 + dz^2). \tag{B.12}$$

The special relativistic line element is invariant under the Lorentz transformation. So is the cosmological line element: it is invariant under the Lorentz-like cosmological transformation. The first keeps invariant the propagation of light, whereas the second keeps invariant the expansion of the Universe. At small velocities with respect to the speed of light, $v \ll c$, the Lorentz transformation goes over to the nonrelativistic Galileo transformation. So is the situation in cosmology: the Lorentz-like cosmological transformation goes over to the nonrelativistic Hubble

Introduction

transformation (see Subsection B.1.1) that is valid for cosmic times much smaller than the Hubble time, $t \ll \tau$.

B.1.3 Five-dimensional manifold of space, time and velocity

If we add the time to the cosmological flat spacevelocity line element, we obtain

$$ds^2 = c^2 dt^2 - (dx^2 + dy^2 + dz^2) + \tau^2 dv^2. \qquad (B.13)$$

Accordingly, we have a five-dimensional manifold of time, space and velocity. The above line element provides a group of transformations O(2,3). At v=const it yields the Minkowskian line element (B.11); at t=const it gives the cosmological line element (B.12); and at a fixed space point, $dx = dy = dz = 0$, it leads to a new two-dimensional line element

$$ds^2 = c^2 dt^2 + \tau^2 dv^2. \qquad (B.14)$$

The groups associated with the above line elements are, of course, O(1,3), O(3,1) and O(2), respectively. They are the Lorentz group, the cosmological group and a two-dimensional Euclidean group, respectively.

In Section B.2 we discuss some properties of the Universe with gravitation in five dimensions. That includes the Bianchi identities, the gravitational field equations, the velocity as an independent coordinate and the energy density in cosmology. In Section B.3 we find the equations of motion of the expanding Universe and show that the Universe is accelerating. In Section B.4 we discuss the important problem of halo dark matter around galaxies by finding the equations of motion of a star moving around a spherically-symmetric galaxy. The equations obtained are *not* Newtonian and instead the Tully-Fisher formula is obtained from our theory. In Section B.5 the cosmological constant is discussed. Our theory predicts that $\Lambda = 3/\tau^2 = 1.934 \times 10^{-35} \text{s}^{-2}$, in agreement with the supernovae experiments teams. In Section B.6 once again we show that the Universe is infinite and open, now by applying redshift analysis, using a new formula that is derived here. Section B.7 is devoted to the concluding remarks. Sections B.8 – B.10 deal with some mathematical conventions.

B.2 Universe with gravitation

The Universe is, of course, not flat but filled up with gravity. When gravitation is invoked, the above spaces become curved Riemannian with the line element

$$ds^2 = g_{\mu\nu} dx^\mu dx^\nu, \qquad (B.15)$$

where μ, ν take the values 0, 1, 2, 3, 4. The coordinates are: $x^0 = ct$; x^1, x^2, x^3 are spatial coordinates; and $x^4 = \tau v$ (the role of the velocity as an independent coordinate will be discussed in the sequel). The signature is $(+ - - - +)$. The metric tensor $g_{\mu\nu}$ is symmetric and thus we have fifteen independent components. They will be a solution of the Einstein field equations in five dimensions. A discussion on the generalization of the Einstein field equations from four to five dimensions will also be given.

The five-dimensional field equations will not explicitly include a cosmological constant. Our cosmological constant, extracted from the theory, will be equal to $\Lambda = 3/\tau^2 = 1.934 \times 10^{-35} \text{s}^{-2}$ (for $H_0 = 70 \text{km/s-Mpc}$). This should be compared with results of the experiments recently done with the supernovae which suggest the value of $\Lambda \approx 10^{-35} \text{s}^{-2}$. Our cosmological constant is derived from the theory itself which is part of the classification of the cosmological spaces to describe deccelerating, constant or accelerating Universe. We now discuss some

Universe with gravitation

basic questions that are encountered in going from four to five dimensions.

B.2.1 The Bianchi identities

The restricted Bianchi identities are given by [2]

$$\left(R^\nu_\mu - \frac{1}{2}\delta^\nu_\mu R\right)_{;\nu} = 0, \qquad (B.16)$$

where $\mu,\nu=0,\ldots,4$. They are valid in five dimensions just as they are in four dimensions. In Eq. (B.16) R^ν_μ and R are the Ricci tensor and scalar, respectively, and a semicolon denotes covariant differentiation. As a consequence we now have five coordinate conditions which permit us to determine five coordinates. For example, one can choose $g_{00} = 1$, $g_{0k} = 0$, $g_{44} = 1$, where $k=1, 2, 3$. These are the co-moving coordinates in five dimensions that keep the clocks and the velocity-measuring instruments synchronized. We will not use these coordinates in this appendix.

B.2.2 The gravitational field equations

In four dimensions these are the Einstein field equations [3]:

$$R_{\mu\nu} - \frac{1}{2}g_{\mu\nu}R = \kappa T_{\mu\nu}, \qquad (B.17)$$

or equivalently

$$R_{\mu\nu} = \kappa\left(T_{\mu\nu} - \frac{1}{2}g_{\mu\nu}T\right), \qquad (B.18)$$

where $T = g_{\alpha\beta}T^{\alpha\beta}$, and we have $R = -\kappa T$. In five dimensions if one chooses Eq. (B.17) as the field equations then Eq. (B.18) is *not* valid (the factor $\frac{1}{2}$ will have to be replaced by $\frac{1}{3}$, and $R = -\frac{2}{3}\kappa T$), and thus there is no symmetry between R and $-\kappa T$.

In fact, if one assumes Eq. (B.18) to be valid in five dimensions then a simple calculation shows that $\kappa T = -\frac{2}{3}R$ and Eq. (B.18) becomes

$$R_{\mu\nu} - \frac{1}{3}g_{\mu\nu}R = \kappa T_{\mu\nu}. \qquad (B.19)$$

Using now the Bianchi identities (B.16) then leads to $\partial R/\partial x^\nu = 0$ and thus $\partial T/\partial x^\nu = 0$. If one chooses, for example, the expression for the energy-momentum tensor to be given by $T^{\mu\nu} = \rho\,(dx^\mu/ds)\,(dx^\nu/ds)$, then $T = T^{\mu\nu}g_{\mu\nu} = \rho$. But if T is a constant then ρ is a constant (independent of time). This is obviously unacceptable situation for the Universe since ρ decreases as the Universe expands. Hence Eq. (B.18) has to be rejected on physical grounds, and the field equations that will be used by us in five dimensions are those given by (B.17).

Finally it is worthwhile mentioning that the only field equations in five dimensions that have symmetry between geometry and matter like those in the Einstein field equations in four dimensions are:

$$R^\nu_\mu - \frac{2}{5}\delta^\nu_\mu R = \kappa T^\nu_\mu, \qquad (B.20)$$

$$R^\nu_\mu = \kappa\left(T^\nu_\mu - \frac{2}{5}\delta^\nu_\mu T\right), \qquad (B.21)$$

with $R = -\kappa T$. While these equations are interesting, however, they do not reduce to Newtonian gravity in the two-body case, and thus they do not seem to be of physical interest.

B.2.3 Velocity as an independent coordinate

First we have to iterate what do we mean by coordinates in general and how one measures them. The time coordinate is measured by clocks as was emphasized by Einstein repeatedly [4,5]. So are the spatial coordinates: they are measured by meters, as

Universe with gravitation 153

was originally done in special relativity theory by Einstein, or by use of Bondi's more modern version of k-calculus [6,7].

But how about the velocity as an independent coordinate? One might incline to think that if we know the spatial coordinates then the velocities are just the time derivative of the coordinates and they are not independent coordinates. This is, indeed, the situation for a dynamical system when the coordinates are given as functions of the time. But in general the situation is different, especially in cosmology. Take, for instance, the Hubble law $v = H_0 x$. Obviously v and x are independent parameters and v is not the time derivative of x. Basically one can measure v by instruments.

B.2.4 Effective mass density in cosmology

To finish this section we discuss the important concept of the energy density in cosmology. We use the Einstein field equations, in which the right-hand side includes the energy-momentum tensor. For fields other than gravitation, like the electromagnetic field, this is a straightforward expression that comes out as a generalization to curved spacetime of the same tensor appearing in special-relativistic electrodynamics. However, when dealing with matter one should construct the energy-momentum tensor according to the physical situation (see, for example, Fock, Ref. 26). Often a special expression for the mass density ρ is taken for the right-hand side of Einstein's equations, which sometimes is expressed as a δ-function.

In cosmology we also have the situation where the mass density is put on the right-hand side of the Einstein field equations. There is also the critical mass density $\rho_c = 3/8\pi G \tau^2$, the value of which is about 10^{-29} g/cm^3, just a few hydrogen atoms per cubic meter throughout the cosmos. If the Universe average mass density ρ is equal to ρ_c then the three spatial geometry of the four-dimensional cosmological space is Euclidean. A deviation from this Euclidean geometry necessitates an increase or decrease from ρ_c. That is to say that

$$\rho_{eff} = \rho - \rho_c \qquad (B.22)$$

is the active or the effective mass density that causes the three geometry not to be Euclidean. Accordingly, one should use ρ_{eff} on the right-hand side of the Einstein field equations. Indeed, we will use such a convention throughout this appendix. The subtraction of ρ_c from ρ in not significant for celestial bodies amd makes no difference.

B.3 The accelerating Universe

B.3.1 Preliminaries

In the last two sections we gave arguments to the fact that the Universe should be presented in five dimensions, even though the standard cosmological theory is obtained from Einstein's four-dimensional general relativity theory. The situation here is similar to that prevailed before the advent of ordinary special relativity. At that time the equations of electrodynamics, written in three dimensions, were well known to predict that the speed of light was constant. But that was not the end of the road. The abandon of the concept of absolute space along with the constancy of the speed of light led to the four-dimensional notion. In cosmology now, we have to give up the notion of absolute cosmic time. Then this with the constancy of the Hubble constant in the limit of zero distance leads us to a five-dimensional presentation of cosmology.

We recall that the field equations are those of Einstein in five dimensions,

$$R_\mu^\nu - \frac{1}{2}\delta_\mu^\nu R = \kappa T_\mu^\nu, \qquad (B.23)$$

where Greek letters $\alpha, \beta, \cdots, \mu, \nu, \cdots = 0, 1, 2, 3, 4$. The coordinates are $x^0 = ct$, x^1, x^2 and x^3 are space-like coordinates, $r^2 = (x^1)^2 + (x^2)^2 + (x^3)^2$, and $x^4 = \tau v$. The metric used is given by (see Section B.8)

$$g_{\mu\nu} = \begin{pmatrix} 1+\phi & 0 & 0 & 0 & 0 \\ 0 & -1 & 0 & 0 & 0 \\ 0 & 0 & -1 & 0 & 0 \\ 0 & 0 & 0 & -1 & 0 \\ 0 & 0 & 0 & 0 & 1+\psi \end{pmatrix}. \qquad (B.24)$$

We will keep only linear terms. The nonvanishing Christoffel symbols are given by (see Section B.8)

$$\Gamma^0_{0\lambda} = \frac{1}{2}\phi_{,\lambda}, \quad \Gamma^0_{44} = -\frac{1}{2}\psi_{,0}, \quad \Gamma^n_{00} = \frac{1}{2}\phi_{,n}, \quad \Gamma^n_{44} = \frac{1}{2}\psi_{,n},$$

$$\Gamma^4_{00} = -\frac{1}{2}\phi_{,4}, \quad \Gamma^4_{4\lambda} = \frac{1}{2}\psi_{,\lambda}, \qquad (B.25)$$

where $n = 1, 2, 3$ and a comma denotes partial differentiation. The components of the Ricci tensor and the Ricci scalar are given by (Section B.9)

$$R^0_0 = \frac{1}{2}\left(\nabla^2\phi - \phi_{,44} - \psi_{,00}\right), \qquad (B.26a)$$

$$R^n_0 = \frac{1}{2}\psi_{,0n}, \quad R^0_n = -\frac{1}{2}\psi_{,0n}, \quad R^4_0 = R^0_4 = 0, \qquad (B.26b)$$

$$R^n_m = \frac{1}{2}\left(\phi_{,mn} + \psi_{,mn}\right), \qquad (B.26c)$$

$$R^4_n = -\frac{1}{2}\phi_{,n4}, \quad R^n_4 = \frac{1}{2}\phi_{,n4}. \qquad (B.26d)$$

$$R^4_4 = \frac{1}{2}\left(\nabla^2\psi - \phi_{,44} - \psi_{,00}\right), \qquad (B.26e)$$

$$R = \nabla^2\phi + \nabla^2\psi - \phi_{,44} - \psi_{,00}. \qquad (B.27)$$

In the above equations ∇^2 is the ordinary three-dimensional Laplace operator.

B.3.2 Expanding Universe

The line element in five dimensions is given by

$$ds^2 = (1+\phi)dt^2 - dr^2 + (1+\psi)dv^2, \qquad (B.28)$$

where $dr^2 = (dx^1)^2 + (dx^2)^2 + (dx^3)^2$, and where c and τ were taken, for brevity, as equal to 1. The line element (B.28) represents a spherically symmetric universe.

The expansion of the universe (the Hubble expansion) is recorded at a definite fixed time and thus $dt = 0$. Accordingly, taking into account $d\theta = d\phi = 0$, Eq. (B.28) gives the following equation for the expansion of the universe at a certain moment,

$$-dr^2 + (1+\psi)dv^2 = 0, \qquad (B.29)$$

and thus

$$\left(\frac{dr}{dv}\right)^2 = 1+\psi. \qquad (B.30)$$

To find ψ we solve the Einstein field equation (noting that $T_0^0 = g_{0\alpha}T^{\alpha 0} \approx T^{00} = \rho(dx^0/ds)^2 \approx c^2\rho$, or $T_0^0 \approx \rho$ in units with $c = 1$):

$$R_0^0 - \frac{1}{2}\delta_0^0 R = 8\pi G \rho_{eff} = 8\pi G (\rho - \rho_c), \qquad (B.31)$$

where $\rho_c = 3/8\pi G\tau^2$.

A simple calculation using Eqs. (B.26a) and (B.27) then yields

$$\nabla^2\psi = 6(1-\Omega), \qquad (B.32)$$

where $\Omega = \rho/\rho_c$.

The solution of the field equation (B.32) is given by

$$\psi = (1-\Omega)r^2 + \psi_0, \qquad (B.33)$$

The accelerating Universe

where the first part on the right-hand side is a solution for the non-homogeneous Eq. (B.32), and ψ_0 represents a solution to its homogeneous part, i.e. $\nabla^2 \psi_0 = 0$. A solution for ψ_0 can be obtained as an infinite series in powers of r. The only term that is left is of the form $\psi_0 = -K_2/r$, where K_2 is a constant whose value can easily be shown to be the Schwarzschild radius, $K_2 = 2GM$. We therefore have

$$\psi = (1-\Omega)r^2 - 2GM/r. \tag{B.34}$$

The universe expansion is therefore given by

$$\left(\frac{dr}{dv}\right)^2 = 1 + (1-\Omega)r^2 - \frac{2GM}{r}. \tag{B.35}$$

For large r the last term on the right-hand side of (B.35) can be neglected, and therefore

$$\left(\frac{dr}{dv}\right)^2 = 1 + (1-\Omega)r^2, \tag{B.36}$$

or

$$\frac{dr}{dv} = \left[1 + (1-\Omega)r^2\right]^{1/2}. \tag{B.37}$$

Inserting now the constants c and τ we finally obtain for the expansion of the universe

$$\frac{dr}{dv} = \tau\left[1 + (1-\Omega)r^2/c^2\tau^2\right]^{1/2}. \tag{B.38}$$

This result is exactly that obtained by Behar and Carmeli (BC) (Eq. (5.10)) when the non-relativistic relation $z = v/c$, where z is the redshift parameter, is inserted in the previous result [8].

The second term in the square brackets of (B.38) represents the deviation from constant expansion due to gravity. For without this term, Eq. (B.38) reduces to $dr/dv = \tau$, thus $r =$

$\tau v + const$. The constant can be taken zero if one assumes, as usual, that at $r = 0$ the velocity should also vanish. Accordingly we have $r = \tau v$ or $v = \tau^{-1} r$. Hence when $\Omega = 1$, namely when $\rho = \rho_c$, we have a constant expansion.

B.3.3 Decelerating, constant and accelerating expansions

The equation of motion (B.38) can be integrated exactly (see Section B.10). The results are:
For the $\Omega > 1$ case

$$r(v) = (c\tau/\alpha) \sin(\alpha v/c); \quad \alpha = (\Omega - 1)^{1/2}. \qquad (B.39)$$

This is obviously a decelerating expansion.
For $\Omega < 1$,

$$r(v) = (c\tau/\beta) \sinh(\beta v/c); \quad \beta = (1 - \Omega)^{1/2}. \qquad (B.40)$$

This is now an accelerating expansion.
For $\Omega = 1$ we have, from Eq. (B.38),

$$d^2 r/dv^2 = 0, \qquad (B.41)$$

whose solution is, of course,

$$r(v) = \tau v, \qquad (B.42)$$

and this is a constant expansion. It will be noted that the last solution can also be obtained directly from the previous two cases for $\Omega > 1$ and $\Omega < 1$ by going to the limit $v \to 0$, using L'Hospital's lemma, showing that our solutions are consistent.

It has been shown in BC that the constant expansion is just a transition stage between the decelerating and the accelerating expansions as the universe evolves toward its present situation. This occurred at 8.5 Gyr ago at a time the cosmic radiation temperature was 143K [8].

B.3.4 Accelerating Universe

In order to decide which of the three cases is the appropriate one at the present time, it will be convenient to write the solutions (B.39), (B.40) and (B.42) in the ordinary Hubble law form $v = H_0 r$. Expanding Eqs. (B.39) and (B.40) and keeping the appropriate terms then yields

$$r = \tau v \left(1 - \alpha^2 v^2 / 6c^2\right), \qquad (B.43)$$

$$r = \tau v \left(1 + \beta^2 v^2 / 6c^2\right), \qquad (B.44)$$

for the $\Omega_M > 1$ and $\Omega_M < 1$ cases, respectively. Using now the expressions for α and β in Eqs. (B.43) and (B.44), then both of the latter can be reduced into the single equation

$$r = \tau v \left[1 + (1 - \Omega_M) v^2 / 6c^2\right]. \qquad (B.45)$$

Inverting now this equation by writing it in the form $v = H_0 r$, we obtain in the lowest approximation for H_0

$$H_0 = h \left[1 - (1 - \Omega_M) v^2 / 6c^2\right], \qquad (B.46)$$

where $h = 1/\tau$. Using $v \approx r/\tau$, or $z \approx v/c$, we also obtain

$$H_0 = h \left[1 - (1 - \Omega_M) r^2 / 6c^2 \tau^2\right] = h \left[1 - (1 - \Omega_M) z^2 / 6\right]. \qquad (B.47)$$

The above equations show that H_0 depends on the distance, or equivalently, on the redshift. Consequently H_0 has meaning only in the limits $r \to 0$ and $z \to 0$, namely when measured *locally*, in which case it becomes the constant h. This is similar to the situation with respect to the speed of light when measured globally in the presence of gravitational field as the ratio between distance and time, the result usually depends on these parameters. Only in the limit one obtains the constant speed of light in vacuum ($c \approx 3 \times 10^{10}$ cm/s).

As is seen from the above discussion, H_0 is intimately related to the sign of the factor $(1 - \Omega_M)$. If measurements of H_0 indicate that it increases with the redshift parameter z then the sign of $(1 - \Omega_M)$ is negative, namely $\Omega_M > 1$. If, however, H_0 decreases when z increases then the sign of $(1 - \Omega_M)$ is positive, i.e. $\Omega_M < 1$. The possibility of H_0 not depends on the redshift parameter indicates that $\Omega_M = 1$. In recent years different measurements were obtained for H_0, with the so-called "short" and "long" distance scales, in which higher values of H_0 were obtained for the short distances and lower values for H_0 corresponded to the long distances [9-18]. Indications are that the longer the distance of measurement, the smaller the value of H_0. If one takes these experimental results seriously, then that is possible only for the case in which $\Omega_M < 1$, namely when the Universe is at an accelerating expansion phase, and the Universe is thus open. We will see in Section B.6 that the same result is obtained via a new cosmological redshift formula.

B.4 The Tully-Fisher formula: Halo dark matter

In this section we derive the equations of motion of a star moving around a spherically symmetric galaxy and show that the Tully-Fisher formula is obtained from our five-dimensional cosmological theory. The calculation is lengthy but it is straightforward. The equations of motion will first be of general nature and only afterward specialized to the motion of a star around the field of a galaxy. The equations obtained are *not* Newtonian. The Tully-Fisher formula was obtained by us in a previous paper [19] using two representations of Einstein's general relativity: the standard spacetime theory and a spacevelocity version of it. However, the present derivation is a straightforward result from

The Tully-Fisher formula: Halo dark matter

the unification of space, time and velocity.

Our notation in this section is as follows: $\alpha, \beta, \gamma, \ldots = 0, \cdots, 4$; $a, b, c, d, \cdots = 0, \cdots, 3$; $p, q, r, s, \cdots = 1, \cdots, 4$; and $k, l, m, n, \cdots = 1, 2, 3$. The coordinates are: $x^0 = ct$ (timelike), $x^k = x^1, x^2, x^3$ (spacelike), and $x^4 = \tau v$ (velocitylike – see Section B.2.3).

B.4.1 The geodesic equation

As usual the equations of motion are obtained in general relativity theory from the covariant conservation law of the energy-momentum tensor (which is a consequence of the restricted Bianchi identities), and the result, as is well known, is the geodesic equation that describes the motion of a spherically symmetric test particle. In our five-dimensional cosmological theory we have five equations of motion. They are given by

$$\frac{d^2 x^\mu}{ds^2} + \Gamma^\mu_{\alpha\beta} \frac{dx^\alpha}{ds} \frac{dx^\beta}{ds} = 0. \qquad (B.48)$$

We now change the independent parameter s into an arbitrary new parameter σ, then the geodesic equation becomes

$$\frac{d^2 x^\mu}{d\sigma^2} + \Gamma^\mu_{\alpha\beta} \frac{dx^\alpha}{d\sigma} \frac{dx^\beta}{d\sigma} = -\frac{d^2\sigma/ds^2}{(d\sigma/ds)^2} \frac{dx^\mu}{d\sigma}. \qquad (B.49)$$

The parameter σ will be taken once as $\sigma = x^0$ (the time coordinate) and then $\sigma = x^4$ (the velocity coordinate). We obtain, for the first case,

$$\frac{d^2 x^\mu}{(dx^0)^2} + \Gamma^\mu_{\alpha\beta} \frac{dx^\alpha}{dx^0} \frac{dx^\beta}{dx^0} = -\frac{d^2 x^0/ds^2}{(dx^0/ds)^2} \frac{dx^\mu}{dx^0}. \qquad (B.50)$$

The right-hand side of Eq. (B.50) can be written in a somewhat different form by using its zero component

$$\frac{d^2 x^0}{(dx^0)^2} + \Gamma^0_{\alpha\beta} \frac{dx^\alpha}{dx^0} \frac{dx^\beta}{dx^0} = -\frac{d^2 x^0/ds^2}{(dx^0/ds)^2} \frac{dx^0}{dx^0}. \qquad (B.51)$$

But $dx^0/dx^0 = 1$, and $d^2x^0/(dx^0)^2 = 0$. Hence we obtain

$$\frac{d^2x^0/ds^2}{(dx^0/ds)^2} = -\Gamma^0_{\alpha\beta}\frac{dx^\alpha}{dx^0}\frac{dx^\beta}{dx^0}. \qquad (B.52)$$

Using the above result in Eq. (B.50), the latter can be written in the form

$$\frac{d^2x^\mu}{(dx^0)^2} + \left(\Gamma^\mu_{\alpha\beta} - \Gamma^0_{\alpha\beta}\frac{dx^\mu}{dx^0}\right)\frac{dx^\alpha}{dx^0}\frac{dx^\beta}{dx^0} = 0. \qquad (B.53)$$

It will be noted that the zero component Eq. (B.53) is now an identity, and consequently it reduces to the four-dimensional equation

$$\frac{d^2x^p}{(dx^0)^2} + \left(\Gamma^p_{\alpha\beta} - \Gamma^0_{\alpha\beta}\frac{dx^p}{dx^0}\right)\frac{dx^\alpha}{dx^0}\frac{dx^\beta}{dx^0} = 0, \qquad (B.54)$$

where $p = 1, 2, 3, 4$.

In exactly the same way we parametrize the geodesic equation (B.48) now with respect to the velocity by choosing the parameter $\sigma = \tau v$. The result is

$$\frac{d^2x^a}{(dx^4)^2} + \left(\Gamma^a_{\alpha\beta} - \Gamma^4_{\alpha\beta}\frac{dx^a}{dx^4}\right)\frac{dx^\alpha}{dx^4}\frac{dx^\beta}{dx^4} = 0, \qquad (B.55)$$

where $a = 0, 1, 2, 3$.

The equation of motion (B.54) will be expanded in terms of the parameter v/c, assuming $v \ll c$, whereas Eq. (B.55) will be expanded with t/τ, where t is a characteristic cosmic time, and $t \ll \tau$. We then can use the Einstein-Infeld-Hoffmann (EIH) method that is well known in general relativity in obtaining the equations of motion [20-37].

We start with Eq. (B.54). As is seen the second term in the parentheses can be neglected with respect to the first one since

The Tully-Fisher formula: Halo dark matter

$d/dx^0 = (1/c)d/dt$, and we obtain

$$\frac{d^2x^p}{(dx^0)^2} + \Gamma^p_{\alpha\beta}\frac{dx^\alpha}{dx^0}\frac{dx^\beta}{dx^0} = 0. \quad (B.56)$$

In Eq. (B.55) we can also neglect the second term in the parentheses since $d/dx^4 = (1/\tau)d/dv$. As a result we have the approximate equations of motion

$$\frac{d^2x^a}{(dx^4)^2} + \Gamma^a_{\alpha\beta}\frac{dx^\alpha}{dx^4}\frac{dx^\beta}{dx^4} = 0. \quad (B.57)$$

The equations of motion (B.56) and (B.57) are consequently given by

$$\frac{d^2x^p}{dt^2} + \Gamma^p_{\alpha\beta}\frac{dx^\alpha}{dt}\frac{dx^\beta}{dt} = 0, \quad (B.58)$$

$$\frac{d^2x^a}{dv^2} + \Gamma^a_{\alpha\beta}\frac{dx^\alpha}{dv}\frac{dx^\beta}{dv} = 0. \quad (B.59)$$

To find the lowest approximation of Eq. (B.58), since $dx^0/dt \gg dx^q/dt$, all terms with indices that are not zero-zero can be neglected. Consequently, Eq. (B.58) is reduced to the form

$$\frac{d^2x^p}{dt^2} \approx -\Gamma^p_{00}, \quad (B.60)$$

in the lowest approximation.

B.4.2 Equations of motion

Accordingly Γ^p_{00} acts like a Newtonian force per mass unit. In terms of the metric tensor we therefore obtain, since $\Gamma^p_{00} = -\frac{1}{2}\eta^{pq}\phi_{,q}$ (see Section B.8)

$$\frac{d^2x^p}{dt^2} \approx -\frac{1}{2}\eta^{pq}\frac{\partial\phi}{\partial q}, \quad (B.61)$$

where $\phi = g_{00} - 1$ (see Section B.8). We now decompose this equation into a spatial ($p = 1, 2, 3$) and a velocity ($p = 4$) parts, getting

$$\frac{d^2 x^k}{dt^2} = -\frac{1}{2}\frac{\partial \phi}{\partial x^k}, \qquad (B.62a)$$

$$\frac{d^2 v}{dt^2} = 0. \qquad (B.62b)$$

Using exactly the same method, Eq. (B.59) yields

$$\frac{d^2 x^k}{dv^2} = -\frac{1}{2}\frac{\partial \psi}{\partial x^k}, \qquad (B.63a)$$

$$\frac{d^2 t}{dv^2} = 0, \qquad (B.63b)$$

where $\psi = g_{44} - 1$. In the above equations $k = 1, 2, 3$. Equation (B.62a) is exactly the law of motion with the function ϕ being twice the Newtonian potential. The other three equations Eq. (B.62b) and Eqs. (B.63a,b) are not Newtonian and are obtained only in the present theory. It remains to find out the functions ϕ and ψ.

To find out the function ϕ we solve the Einstein field equation (noting that $T_4^4 = g_{4\alpha} T^{\alpha 4} \approx T^{44} = \rho (dx^4/ds)^2 \approx \tau^2 \rho$, and thus $T_4^4 \approx \rho$ in units in which $\tau = 1$):

$$R_4^4 - \frac{1}{2}\delta_4^4 R = 8\pi G \rho_{eff} = 8\pi G (\rho - \rho_c). \qquad (B.64)$$

A straightforward calculation then gives

$$\nabla^2 \phi = 6 \left(1 - \Omega_M\right), \qquad (B.65)$$

whose solution is given by

$$\phi = (1 - \Omega_M) r^2 + \phi_0, \qquad (B.66)$$

where ϕ_0 is a solution of the homogeneous equation $\nabla^2\phi_0 = 0$. One then easily finds that $\phi_0 = -K_1/r$, where $K_1 = 2GM$. In the same way the function ψ can be found (see Section B.3),

$$\psi = (1 - \Omega_M)r^2 + \psi_0, \qquad (B.67)$$

with $\nabla^2\psi_0 = 0$, $\psi_0 = -K_2/r$ and $K_2 = 2GM$. (When units are inserted then $K_1 = 2GM/c^2$ and $K_2 = 2GM\tau^2/c^2$.) For the purpose of obtaining equations of motion one can neglect the terms $(1 - \Omega_M)r^2$, actually $(1 - \Omega_M)r^2/c^2\tau^2$, in the solutions for ϕ and ψ. One then obtains

$$g_{00} \approx 1 - 2GM/c^2 r, \quad g_{44} \approx 1 - 2GM\tau^2/c^2 r. \qquad (B.68)$$

The equations of motion (B.62a) and (B.63a), consequently, have the forms

$$\frac{d^2 x^k}{dt^2} = \left(\frac{GM}{r}\right)_{,k}, \quad \frac{d^2 x^k}{dv^2} = \left(\frac{GM}{r}\right)_{,k}, \qquad (B.69)$$

or, when inserting the constants c and τ,

$$\frac{d^2 x^k}{dt^2} = GM\left(\frac{1}{r}\right)_{,k}, \qquad (B.70a)$$

$$\frac{d^2 x^k}{dv^2} = kM\left(\frac{1}{r}\right)_{,k}, \qquad (B.70b)$$

where $k = G\tau^2/c^2$. It remains to integrate equations (B.62b) and (B.63b). One finds that $v = a_0 t$, where a_0 is a constant which can be taken as equal to $a_0 = c/\tau \approx cH_0$. Accordingly, we see that the particle experiences an acceleration $a_0 = c/\tau \approx cH_0$ directed outward when the motion is circular.

Equation (B.70a) is Newtonian but (B.70b) is not. The integration of the latter is identical to that familiar in classical Newtonian mechanics, but there is an essential difference which should be emphasized. In Newtonian equations of motion one deals with a path of motion in the 3-space. In our theory we do not have that situation. Rather, the paths here indicate locations of particles in the sense of the Hubble distribution, which now takes a different physical meaning. With that in mind we proceed as follows.

Equation (B.70b) yields the first integral

$$\left(\frac{ds}{dv}\right)^2 = \frac{kM}{r}, \qquad (B.71a)$$

where v is the velocity of the particles, in analogy to the Newtonian

$$\left(\frac{ds}{dt}\right)^2 = \frac{GM}{r}. \qquad (B.71b)$$

In these equations s is the length parameter along the path of the accumulation of the particles.

Comparing Eqs. (B.71a) and (B.71b), we obtain

$$\frac{ds}{dv} = \frac{\tau}{c}\frac{ds}{dt}. \qquad (B.72)$$

Thus

$$\frac{dv}{dt} = \frac{c}{\tau}. \qquad (B.73)$$

Accordingly, as we have mentioned before, the particle experiences an acceleration $a_0 = c/\tau \approx cH_0$ directed outward when the motion is circular.

B.4.3 The Tully-Fisher law

The motion of a particle in a central field is best described in terms of an "effective potential", V_{eff}. In Newtonian mechanics this is given by [38]

$$V_{eff} = -\frac{GM}{r} + \frac{L^2}{2r^2}, \qquad (B.74)$$

where L is the angular momentum per mass unit. In our case the effective potential is

$$V_{eff}(r) = -\frac{GM}{r} + \frac{L^2}{2r^2} + a_0 r. \qquad (B.75)$$

The circular motion is obtained at the minimal value of (B.75), i.e.,

$$\frac{dV_{eff}}{dr} = \frac{GM}{r^2} - \frac{L^2}{r^3} + a_0 = 0, \qquad (B.76)$$

with $L = v_c r$, and v_c is the circular velocity. This gives

$$v_c^2 = \frac{GM}{r} + a_0 r. \qquad (B.77)$$

Thus

$$v_c^4 = \left(\frac{GM}{r}\right)^2 + 2GM a_0 + a_0^2 r^2, \qquad (B.78)$$

where $a_0 = c/\tau \approx cH_0$.

The first term on the right-hand side of Eq. (B.78) is purely Newtonian, and cannot be avoided by any reasonable theory. The second one is the Tully-Fisher term. The third term is extremely small at the range of distances of stars around a galaxy. It is well known that astronomical observations show that for disk galaxies the fourth power of the circular velocity of stars moving around the core of the galaxy, v_c^4, is proportional to the total luminosity L of the galaxy to an accuracy of more than two orders of magnitude in L, namely $v_c^4 \propto L$ [39]. Since L is proportional to the mass M of the galaxy, one obtains $v_c^4 \propto M$. This is the Tully-Fisher law. There is no dependence on the distance of the star from the center of the galaxy as Newton's law $v_c^2 = GM/r$ requires for circular motion. In order to rectify this deviation from Newton's laws, astronomers assume the existence of halos around the galaxy which are filled with dark matter and arranged in such a way so as to satisfy the Tully-Fisher law for each particular situation.

In conclusion it appears that there is no necessity for the assumption of the existence of halo dark matter around galaxies. Rather, the result can be described in terms of the properties of spacetimevelocity.

B.5 The cosmological constant

B.5.1 The cosmological term

First, a historical remark. In order to allow the existence of a static solution for the gravitational field equations, Einstein made a modification to his original equations (B.17) by adding a cosmological term,

$$R_{\mu\nu} - \frac{1}{2}g_{\mu\nu}R + \Lambda g_{\mu\nu} = \kappa T_{\mu\nu}, \quad (B.79)$$

where Λ is the cosmological constant and $\kappa = 8\pi G$ (c is taken as unity). For a homogeneous and isotropic Universe with the line element [40,41]

$$ds^2 = dt^2 - a^2(t) R_0^2 \left[\frac{dr^2}{1 - kr^2} + r^2 \left(d\theta^2 + \sin^2\theta d\phi^2 \right) \right], \quad (B.80)$$

where k is the *curvature parameter* ($k = 1, 0, -1$) and $a(t) = R(t)/R_0$ is the scale factor, with the energy-momentum tensor

$$T_{\mu\nu} = (\rho + p) u_\mu u_\nu - p g_{\mu\nu}, \quad (B.81)$$

Einstein's equations (B.79) reduce to the two Friedmann equations

$$H^2 \equiv \left(\frac{\dot{a}}{a}\right)^2 = \frac{\kappa}{3}\rho + \frac{\Lambda}{3} - \frac{k}{a^2 R_0^2}, \quad (B.82)$$

The cosmological constant 169

$$\frac{\ddot{a}}{a} = -\frac{\kappa}{6}(\rho + 3p) + \frac{\Lambda}{3}. \qquad (B.83)$$

These equations admit a static solution ($\dot{a} = 0$) with $k > 0$ and $\Lambda > 0$.

After Hubble's discovery that the Universe is expanding, the role of the cosmological constant to allow static homogeneous solutions to Einstein's equations in the presence of matter, seemed to be unnecessary. For a long time the cosmological term was considered to be of no physical interest in cosmological problems.

From the Friedmann equation (B.82), for any value of the Hubble parameter H there is a critical value of the mass density such that the spatial geometry is flat ($k = 0$), $\rho_c = 3H_0^2/\kappa$ (see Subsection B.2.4). One usually measures the total mass density in terms of the critical density ρ_c by means of the density parameter $\Omega_M = \rho/\rho_c$.

In general, the mass density ρ includes contributions from various distinct components. From the point of view of cosmology, the relevant aspect of each component is how its energy density evolves as the Universe expands. In general, a positive Λ causes acceleration to the Universe expansion, whereas a negative Λ and ordinary matter tend to decelerate it. Moreover, the relative contributions of the components to the energy density change with time. For $\Omega_\Lambda < 0$, the Universe will always recollapse to a Big Crunch. For $\Omega_\Lambda > 0$ the Universe will expand forever unless there is sufficient matter to cause recollapse before Ω_Λ becomes dynamically important. For $\Omega_\Lambda = 0$ we have the familiar situation in which $\Omega_M \leq 1$ Universes expand forever and $\Omega_M > 1$ Universes recollapse. (For more details see the paper by Behar and Carmeli, Ref. 8.)

B.5.2 The supernovae experiments value for the cosmological constant

Recently two groups, the *Supernova Cosmology Project Collaboration* and the *High-Z Supernova Team Collaboration*, presented evidence that the expansion of the Universe is accelerating [42-48]. These teams have measured the distances to cosmological supernovae by using the fact that the intrinsic luminosity of Type Ia supernovae is closely correlated with their decline rate from maximum brightness, which can be independently measured. These measurements, combined with redshift data for the supernovae, led to the prediction of an accelerating Universe. Both teams obtained

$$\Omega_M \approx 0.3, \quad \Omega_\Lambda \approx 0.7, \qquad (B.84)$$

and strongly ruled out the traditional $(\Omega_M, \Omega_\Lambda)=(1, 0)$ Universe. This value of the density parameter Ω_Λ corresponds to a cosmological constant that is small but nonzero and positive,

$$\Lambda \approx 10^{-52} \mathrm{m}^{-2} \approx 10^{-35} \mathrm{s}^{-2}. \qquad (B.85)$$

B.5.3 The Behar-Carmeli predicted value for the cosmological constant

Behar and Carmeli presented a four-dimensional cosmological relativity theory that unifies space and velocity that predicts the acceleration of the Universe and hence it is tacitly having a positive value for Λ in it. As is well known, in the traditional work of Friedmann when added to it a cosmological constant, the field equations obtained are highly complicated and no solutions have been obtained so far. Behar-Carmeli's theory, on the other hand, yields exact solutions and describes the Universe as having a three-phase evolution with a decelerating expansion followed

The cosmological constant 171

by a constant and an accelerating expansion, and it predicts that the Universe is now in the latter phase. In the framework of this theory the zero-zero component of Einstein's equations is written as

$$R_0^0 - \frac{1}{2}\delta_0^0 R = \kappa \rho_{eff} = \kappa\left(\rho - \rho_c\right), \qquad (B.86)$$

where $\rho_c = 3/\kappa\tau^2 \approx 3H^2/\kappa$ is the critical mass density. Comparing Eq. (B.86) with the zero-zero component of Eq. (B.79), one obtains the expression for the cosmological constant in the Behar-Carmeli theory,

$$\Lambda = \kappa\rho_c = 3/\tau^2 \approx 3H^2. \qquad (B.87)$$

B.5.4 Comparison with experiment

To find out the numerical value of τ we use the relationship between $h = \tau^{-1}$ and H_0 given by Eq. (B.47) (CR denote values according to Cosmological Relativity):

$$H_0 = h\left[1 - \left(1 - \Omega_M^{CR}\right)z^2/6\right], \qquad (B.88)$$

where $z = v/c$ is the redshift and $\Omega_M^{CR} = \rho_M/\rho_c$ with $\rho_c = 3h^2/8\pi G$. (Notice that our $\rho_c = 1.194 \times 10^{-29} \text{g/cm}^3$ is different from the standard ρ_c defined with H_0.) The redshift parameter z determines the distance at which H_0 is measured. We choose $z = 1$ and take for

$$\Omega_M^{CR} = 0.245, \qquad (B.89)$$

its value at the present time (corresponds to 0.32 in the standard theory), Eq. (B.88) then gives

$$H_0 = 0.874h. \qquad (B.90)$$

At the value $z = 1$ the corresponding Hubble parameter H_0 according to the latest results from HST can be taken [9] as

$H_0 = 70$km/s-Mpc, thus $h = (70/0.874)$km/s-Mpc, or

$$h = 80.092 \text{km/s-Mpc}, \qquad (B.91)$$

and

$$\tau = 12.486 \text{Gyr} = 3.938 \times 10^{17} \text{s}. \qquad (B.92)$$

What is left is to find the value of Ω_Λ^{CR}. We have $\Omega_\Lambda^{CR} = \rho_c^{ST}/\rho_c$, where $\rho_c^{ST} = 3H_0^2/8\pi G$ and $\rho_c = 3h^2/8\pi G$. Thus $\Omega_\Lambda^{CR} = (H_0/h)^2 = 0.874^2$, or

$$\Omega_\Lambda^{CR} = 0.764. \qquad (B.93)$$

As is seen from Eqs. (B.89) and (B.93) one has

$$\Omega_T = \Omega_M^{CR} + \Omega_\Lambda^{CR} = 0.245 + 0.764 = 1.009 \approx 1, \qquad (B.94)$$

which means the Universe is Euclidean.

As a final result we calculate the cosmological constant according to Eq. (B.87). One obtains

$$\Lambda = 3/\tau^2 = 1.934 \times 10^{-35} \text{s}^{-2}. \qquad (B.95)$$

These results confirm those of the supernovae experiments and indicate the existence of the dark energy as has recently received confirmation from the Boomerang cosmic microwave background experiment [50,51], which showed that the Universe is Euclidean.

B.6 Cosmological redshift analysis

B.6.1 The redshift formula

In this section we derive a general formula for the redshift in which the term $(1 - \Omega_M)$ appears explicitly. Since there are enough data of measurements of redshifts, this allows one to

Cosmological redshift analysis

determine what is the sign of $(1-\Omega_M)$, positive, zero or negative. Our conclusion is that $(1-\Omega_M)$ cannot be negative or zero. This means that the Universe is infinite, and expands forever, a result favored by some cosmologists [49]. To this end we proceed as follows.

Having the metric tensor from Section B.4 we may now find the redshift of light emitted in the cosmos. As usual, at two points 1 and 2 we have for the wave lengths and frequencies:

$$\frac{\lambda_2}{\lambda_1} = \frac{\nu_1}{\nu_2} = \frac{ds\,(2)}{ds\,(1)} = \sqrt{\frac{g_{00}\,(2)}{g_{00}\,(1)}}. \qquad (B.96)$$

Using now the solution for $g_{00} = 1 + \phi$, with ϕ given by Eq. (B.66), in Eq. (B.96), we obtain

$$\frac{\lambda_2}{\lambda_1} = \sqrt{\frac{1 + r_2^2/a^2 - R_s/r_2}{1 + r_1^2/a^2 - R_s/r_1}}. \qquad (B.97)$$

In Eq. (B.97) $R_s = 2GM/c^2$ and $a^2 = c^2\tau^2/(1-\Omega_M)$.

For a sun-like body with radius R located at the coordinates origin, and an observer at a distance r from the center of the body, we then have $r_2 = r$ and $r_1 = R$, thus

$$\frac{\lambda_2}{\lambda_1} = \sqrt{\frac{1 + r^2/a^2 - R_s/r}{1 + R^2/a^2 - R_s/R}}. \qquad (B.98)$$

B.6.2 Particular cases

Since $R \ll r$ and $R_s < R$ is usually the case we can write, to a good approximation,

$$\frac{\lambda_2}{\lambda_1} = \sqrt{\frac{1 + r^2/a^2}{1 - R_s/R}}. \qquad (B.99)$$

The term r^2/a^2 in Eq. (B.99) is a pure cosmological one, whereas R_s/R is the standard general relativistic term. For $R \gg R_s$ we then have

$$\frac{\lambda_2}{\lambda_1} = \sqrt{1 + \frac{r^2}{a^2}} = \sqrt{1 + \frac{(1-\Omega_M)\,r^2}{c^2\tau^2}} \qquad (B.100)$$

for the pure cosmological contribution to the redshift. If, furthermore, $r \ll a$ we then have

$$\frac{\lambda_2}{\lambda_1} = 1 + \frac{r^2}{2a^2} = 1 + \frac{(1-\Omega_M)\,r^2}{2c^2\tau^2} \qquad (B.101)$$

to the lowest approximation in r^2/a^2, and thus

$$z = \frac{\lambda_2}{\lambda_1} - 1 = \frac{r^2}{2a^2} = \frac{(1-\Omega_M)\,r^2}{2c^2\tau^2}. \qquad (B.102)$$

When the contribution of the cosmological term r^2/a^2 is negligible, we then have

$$\frac{\lambda_2}{\lambda_1} = \frac{1}{\sqrt{1 - R_s/R}}. \qquad (B.103)$$

The redshift could then be very large if R, the radius of the emitting body, is just a bit larger than the Schwarzschild radius R_s. For example if $R_s/R = 0.96$ the redshift $z = 4$. For a typical sun like ours, $R_s \ll R$ and we can expand the right-hand side of Eq. (B.103), getting

$$\frac{\lambda_2}{\lambda_1} = 1 + \frac{R_s}{2R}, \qquad (B.104)$$

thus

$$z = \frac{R_s}{2R} = \frac{Gm}{c^2 R}, \qquad (B.105)$$

the standard general relativistic result.

From Eqs. (B.100)–(B.102) it is clear that Ω_M cannot be larger than one since otherwise z will be negative, which means blueshift, and as is well known nobody sees such a thing. If $\Omega_M = 1$ then $z = 0$, and for $\Omega_M < 1$ we have $z > 0$. The case of $\Omega_M = 1$ is also implausible since the light from stars we see is usually redshifted more than the redshift due to the gravity of the body emitting the radiation, as is evident from our sun, for example, whose emitted light is shifted by only $z = 2.12 \times 10^{-16}$ [3].

B.6.3 Conclusions

One can thus conclude that the theory of unified space, time and velocity predicts that the Universe is open. As is well known the standard FRW model does not relate the cosmological redshift to the kind of Universe.

B.7 Concluding remarks

The most direct evidence that the Universe expansion is accelerating and propelled by "dark energy", is provided by the faintness of Type Ia supernovae (SNe Ia) at $z \approx 0.5$ [42,46]. Beyond the redshift range of $0.5 < z < 1$, the Universe was more compact and the attraction of matter was stronger than the repulsive dark energy. At $z > 1$ the expansion of the Universe should have been decelerating [52]. At $z \geq 1$ one would expect an apparent brightness increase of SNe Ia relative to what is supposed to be for a non-decelerating Universe [53].

More confirmation to the Universe accelerating expansion came from the most distant supernova, SN 1997ff, that was recorded by the Hubble Space Telescope. As has been pointed out before, if we look back far enough, we should find a decelerating expansion. Beyond $z = 1$ one should see an earlier time when the mass density was dominant. The measurements obtained from SN 1997ff's redshift and brightness provide a direct proof for the transition from past decelerating to present accelerating expansion. The measurements also exclude the possibility that the acceleration of the Universe is not real but is due to other astrophysical effects such as dust.

B.8 Mathematical conventions and Christoffel symbols

Throughout this appendix we use the convention

$$\alpha, \beta, \gamma, \delta, \cdots = 0, 1, 2, 3, 4,$$

$$a, b, c, d, \cdots = 0, 1, 2, 3,$$

$$p, q, r, s, \cdots = 1, 2, 3, 4,$$

$$k, l, m, n, \cdots = 1, 2, 3.$$

The coordinates are $x^0 = ct$, x^1, x^2 and x^3 are space-like coordinates, $r^2 = (x^1)^2 + (x^2)^2 + (x^3)^2$, and $x^4 = \tau v$. The signature is $(+ - - - +)$. The metric, approximated up to ϕ and ψ, is:

$$g_{\mu\nu} = \begin{pmatrix} 1+\phi & 0 & 0 & 0 & 0 \\ 0 & -1 & 0 & 0 & 0 \\ 0 & 0 & -1 & 0 & 0 \\ 0 & 0 & 0 & -1 & 0 \\ 0 & 0 & 0 & 0 & 1+\psi \end{pmatrix}, \qquad (B.106)$$

$$g^{\mu\nu} = \begin{pmatrix} 1-\phi & 0 & 0 & 0 & 0 \\ 0 & -1 & 0 & 0 & 0 \\ 0 & 0 & -1 & 0 & 0 \\ 0 & 0 & 0 & -1 & 0 \\ 0 & 0 & 0 & 0 & 1-\psi \end{pmatrix}. \qquad (B.107)$$

The nonvanishing Christoffel symbols are (in the linear approximation):

$$\Gamma^0_{0\lambda} = \frac{1}{2}\phi_{,\lambda}, \quad \Gamma^0_{44} = -\frac{1}{2}\psi_{,0}, \quad \Gamma^n_{00} = \frac{1}{2}\phi_{,n}, \qquad (B.108a)$$

$$\Gamma^n_{44} = \frac{1}{2}\psi_{,n}, \quad \Gamma^4_{00} = -\frac{1}{2}\phi_{,4}, \quad \Gamma^4_{4\lambda} = \frac{1}{2}\psi_{,\lambda}, \qquad (B.108b)$$

$$\Gamma^a_{00} = -\frac{1}{2}\eta^{ab}\phi_{,b}, \quad \Gamma^a_{44} = -\frac{1}{2}\eta^{ab}\psi_{,b}, \qquad (B.108c)$$

$$\Gamma^p_{00} = -\frac{1}{2}\eta^{pq}\phi_{,q}, \quad \Gamma^p_{44} = -\frac{1}{2}\eta^{pq}\psi_{,q}. \qquad (B.108d)$$

The Minkowskian metric η in five dimensions is given by

$$\eta = \begin{pmatrix} 1 & 0 & 0 & 0 & 0 \\ 0 & -1 & 0 & 0 & 0 \\ 0 & 0 & -1 & 0 & 0 \\ 0 & 0 & 0 & -1 & 0 \\ 0 & 0 & 0 & 0 & 1 \end{pmatrix}. \qquad (B.109)$$

B.9 Components of the Ricci tensor

The elements of the Ricci tensor are:

$$R_{00} = \frac{1}{2}\left(\nabla^2\phi - \phi_{,44} - \psi_{,00}\right), \qquad (B.110)$$

$$R_{0n} = -\frac{1}{2}\psi_{,0n}, \quad R_{04} = 0, \qquad (B.111)$$

$$R_{mn} = -\frac{1}{2}\left(\phi_{,mn} + \psi_{,mn}\right), \qquad (B.112)$$

$$R_{4n} = -\frac{1}{2}\phi_{,4n}, \qquad (B.113)$$

$$R_{44} = \frac{1}{2}\left(\nabla^2\psi - \phi_{,44} - \psi_{,00}\right). \qquad (B.114)$$

Appendix B. Five-Dimensional Brane World Theory

The Ricci scalar is

$$R = \nabla^2\phi + \nabla^2\psi - \phi_{,44} - \psi_{,00}. \quad (B.115)$$

The mixed Ricci tensor is given by

$$R^0_0 = \frac{1}{2}\left(\nabla^2\phi - \phi_{,44} - \psi_{,00}\right), \quad (B.116)$$

$$R^n_0 = \frac{1}{2}\psi_{,0n}, \quad R^0_n = -\frac{1}{2}\psi_{,0n}, \quad (B.117)$$

$$R^4_0 = R^0_4 = 0, \quad (B.118)$$

$$R^n_m = \frac{1}{2}\left(\phi_{,mn} + \psi_{,mn}\right), \quad (B.119)$$

$$R^4_n = -\frac{1}{2}\phi_{,n4}, \quad R^n_4 = \frac{1}{2}\phi_{,n4}, \quad (B.120)$$

$$R^4_4 = \frac{1}{2}\left(\nabla^2\psi - \phi_{,44} - \psi_{,00}\right). \quad (B.121)$$

B.10 Integration of the Universe expansion equation

The Universe expansion was shown to be given by Eq. (B.38),

$$\frac{dr}{dv} = \tau\left[1 + (1 - \Omega_M)\,r^2/c^2\tau^2\right]^{1/2}.$$

This equation can be integrated exactly by the substitutions

$$\sin\chi = \alpha r/c\tau; \quad \Omega_M > 1 \quad (B.122a)$$

$$\sinh\chi = \beta r/c\tau; \quad \Omega_M < 1 \quad (B.122b)$$

where

$$\alpha = (\Omega_M - 1)^{1/2}, \quad \beta = (1 - \Omega_M)^{1/2}. \quad (B.123)$$

Integration of the Universe expansion equation

For the $\Omega_M > 1$ case a straightforward calculation using Eq. (B.122a) gives

$$dr = (c\tau/\alpha)\cos\chi\, d\chi \qquad (B.124)$$

and the equation of the Universe expansion (B.38) yields

$$d\chi = (\alpha/c)\, dv. \qquad (B.125a)$$

The integration of this equation gives

$$\chi = (\alpha/c)\, v + \text{const.} \qquad (B.126a)$$

The constant can be determined using Eq. (B.122a). At $\chi = 0$, we have $r = 0$ and $v = 0$, thus

$$\chi = (\alpha/c)\, v, \qquad (B.127a)$$

or, in terms of the distance, using (B.122a) again,

$$r(v) = (c\tau/\alpha)\sin\alpha v/c; \qquad \alpha = (\Omega_M - 1)^{1/2}. \qquad (B.128a)$$

This is obviously a decelerating expansion.

For $\Omega_M < 1$, using Eq. (B.122b), a similar calculation yields for the Universe expansion (B.38)

$$d\chi = (\beta/c)\, dv, \qquad (B.125b)$$

thus

$$\chi = (\beta/c)\, v + \text{const.} \qquad (B.126b)$$

Using the same initial conditions as above then gives

$$\chi = (\beta/c)\, v \qquad (B.127b)$$

and in terms of distances,

$$r(v) = (c\tau/\beta)\sinh\beta v/c; \qquad \beta = (1 - \Omega_M)^{1/2}. \qquad (B.128b)$$

180   *Appendix B. Five-Dimensional Brane World Theory*

This is now an accelerating expansion.

For $\Omega_M = 1$ we have, from Eq. (B.38),

$$d^2r/dv^2 = 0. \qquad (B.125c)$$

The solution is, of course,

$$r(v) = \tau v. \qquad (B.128c)$$

This is a constant expansion.

B.11 References

1. M. Carmeli, *Cosmological Special Relativity: The Large-Scale Structure of Space, Time and Velocity* (World Scientific, River Edge, NJ and Singapore, 1997).
2. L.P. Eisenhart, *Riemannian Geometry* (Princeton University Press, Princeton, NJ, 1949).
3. M. Carmeli, *Classical Fields: General Relativity and Gauge Theory* (John Wiley, New York, 1982).
4. A. Einstein, *Autobiographical Notes*, Ed. P.A. Schilpp (Open Court Pub. Co., La Salle and Chicago, 1979).
5. A. Einstein, "Zur Electrodynamik bewegter Körper," *Annalen der Physik* **17**, 891-921 (1905); English translation:"On the electrodynamics of moving bodies," in: A. Einstein, H.A. Lorentz, H. Minkowski and H. Weyl, *The Principle of Relativity* (Dover Publications, 1923), pp. 35-65.
6. H. Bondi, Some special solutions of the Einstein equations, in: *Lectures on General Relativity.* (Prentice-Hall, Inc., Englewood Cliffs, NJ, 1965.) (Brandeis Summer Institute in Theoretical Physics, Vol. 1, 1964.)
7. H. Bondi, *Relativity and Common Sense: A New Approach to Einstein* (Dover Publications, Inc., New York, 1962).
8. S. Behar and M. Carmeli, *Intern. J. Theor. Phys.* **39**, 1375

(2000) (astro-ph/0008352).

9. W.L. Freedman et al., Final results from the Hubble Space Telescope, Talk given at the 20th Texas Symposium on Relativistic Astrophysics, Austin, Texas 10-15 December 2000. (astro-ph/0012376)

10. W.L. Freedman et al., *Nature* **371**, 757 (1994).

11. M. Pierce et al., *Nature* **371**, 385 (1994).

12. B. Schmidt et al., *Astrophys. J.* **432**, 42 (1994).

13. A. Riess et al., *Astrophys. J.* **438**, L17 (1995).

14. A. Sandage et al., *Astrophys. J.* **401**, L7 (1992).

15. D. Branch, *Astrophys. J.* **392**, 35 (1992).

16. B. Schmidt et al., *Astrophys. J.* **395**, 366 (1992).

17. A. Saha et al., *Astrophys. J.* **438**, 8 (1995).

18. P.J.E. Peebles, Status of the big bang cosmology, p. 84, in: *Texas/Pascos 92: Relativistic Astrophysics and Particle Cosmology*, Eds. C.W. Akerlof and M.A. Srednicki, Vol. 688 (The New York Academy of Sciences, New York, 1993).

19. M. Carmeli, *Intern. J. Theor. Phys.* **39**, 1397 (2000) (astro-ph/9907244).

20. A. Einstein and J. Grommer, *Preuss. Akad. Wiss., Phys.-Math. Klasse* **1**, 2 (1927).

21. A. Einstein, L. Infeld and B. Hoffmann, *Annals of Mathematics* **39**, 65 (1938).

22. A. Einstein and L. Infeld, *Can. J. Math.* **1**, 209 (1949).

23. L. Infeld and A. Schild, *Revs. Mod. Phys.* **21**, 408 (1949).

24. L. Infeld, *Revs. Mod. Phys.* **29**, 398 (1957).

25. V. Fock, *Revs. Mod. Phys.* **29**, 325 (1957).

26. V. Fock, *The Theory if Space, Time and Gravitation* (Pergamen Press, Oxford, 1959).

27. L. Infeld and J. Plebanski, *Motion and Relativity* (Pergamen Press, Oxford, 1960).

28. B. Bertotti and J. Plebanski, *Ann. Phys. (N.Y.)* **11**, 169 (1960).

29. M. Carmeli, *Phys. Lett.* **9**, 132 (1964).
30. M. Carmeli, *Phys. Lett.* **11**, 24 (1964).
31. M. Carmeli, *Ann. Phys. (N.Y.)* **30**, 168 (1964).
32. M. Carmeli, *Phys. Rev.* **138**, B1003 (1965).
33. M. Carmeli, *Nuovo Cimento* **37**, 842 (1965).
34. M. Carmeli, *Ann. Phys. (N.Y.)* **34**, 465 (1965).
35. M. Carmeli, *Ann. Phys. (N.Y.)* **35**, 250 (1965).
36. M. Carmeli, *Phys. Rev.* **140**, B1441 (1965).
37. T. Damour, Gravitational radiation and the motion of compact bodies, in *Gravitational Radiation*, N. Deruelle and T. Piran, Eds. (North-Holland, Amsterdam 1983), pp. 59-144.
38. H. Goldstein, *Classical Mechanics* (Addison-Wesley, Reading, MA, 1980).
39. B.C. Whitemore, Rotation curves of spiral galaxies in clusters, in: *Galactic Models*, J.R. Buchler, S.T. Gottesman, J.H. Hunter. Jr., Eds., (New York Academy Sciences, New York, 1990).
40. H.C. Ohanian and R. Ruffini, *Gravitation and Spacetime*, Second Edition (W.W. Norton, New York and London, 1994).
41. L.D. Landau and E.M. Lifshitz, *The Classical Theory of Fields* (Pergamon Press, Oxford, 1979).
42. P.M. Garnavich *et al.*, *Astrophys. J.* **493**, L53 (1998). [Hi-Z Supernova Team Collaboration (astro-ph/9710123)].
43. B.P. Schmidt *et al.*, *Astrophys. J.* **507**, 46 (1998). [Hi-Z Supernova Team Collaboration (astro-ph/9805200)].
44. A.G. Riess *et al.*, *Astronom. J.* **116**, 1009 (1998). [Hi-Z Supernova Team Collaboration (astro-ph/9805201)].
45. P.M. Garnavich *et al.*, *Astrophys. J.* **509**, 74 (1998). [Hi-Z Supernova Team Collaboration (astro-ph/9806396)].
46. S. Perlmutter *et al.*, *Astrophys. J.* **483**, 565 (1997). [Supernova Cosmology Project Collaboration (astro-ph/9608192)].
47. S. Perlmutter *et al.*, *Nature* **391**, 51 (1998). [Supernova Cosmology Project Collaboration (astro-ph/9712212)].

48. S. Perlmutter *et al.*, *Astrophys. J.* **517**, 565 (1999). [Supernova Cosmology Project Collaboration (astro-ph/9812133)].
49. M.S. Turner, *Science* **262**, 861 (1993).
50. P. de Bernardis *et al.*, *Nature* **404**, 955 (2000). (astro-ph/0004404)
51. J.R. Bond *et al.*, in *Proc. IAU Symposium* 201 (2000). (astro-ph/0011378)
52. A.V. Filippenko and A.G. Riess, p.227 in: *Particle Physics and Cosmology: Second Tropical Workshop*, J.F. Nieves, Editor (AIP, New York, 2000).
53. A.G. Riess *et al.*, *Astrophys. J.*, in press. (astro-ph/0104455)

Appendix C

Cosmic Temperature Decline

In this appendix we discuss the early stage of the Universe after the Big Bang at a time where no stars or galaxies existed but only a uniform hot plasma of free electrons and nuclei. The Universe temperature was determined by the Stefan-Boltzmann law of thermodynamics and the general relativistic cosmological theory. At the present time one has the background cosmic radiation with the temperature of 2.73K. We calculate how much of the early Universe energy has gone to matter and other forms of energy, so as to leave us with a background radiation of only 2.73K [1].

C.1 Introduction

At the early stage of the Universe-evolution there were no stars and no galaxies, but only a uniform hot plasma consisting of free electrons and free nuclei. At very early times, the violent thermal collisions would have prevented the existence of any kind of nucleus, and the matter in the Universe must have been in the form of free electrons, protons and neutrons. The temperature of the Universe is related to its evolution and, at the early stage, is given by a well-known formula which shows that $T \propto t^{-1/2}$. The question raised here is how to relate this temperature at the very early stage of the Universe to the present time cosmic background temperature 2.73K. Obviously part of the energy that the Universe had at the early stage has gone to the matter, and the background radiation temperature represents the other part. In the following it is shown that the ratio of energy that goes to matter to that of the background radiation is about 13. In other words, if the Universe today would have no matter, the background cosmic temperature would be about 35K (13×2.73K).

C.2 Temperature formula without gravity

Our starting point is the familiar thermodynamical formula that relates the temperature to the cosmic time with respect to the Big Bang [2,3]:

$$T = \left(\frac{45\hbar^3}{32\pi^3 k^4 G}\right)^{1/4} t^{-1/2}. \qquad (C.1)$$

In this equation k is Boltzmann's constant and G is Newton's gravitational constant. Our aim is to transform this temperature to the present time. This can be done by using the cosmological transformation [1]

$$T = \frac{T_0}{\left(1 - \tilde{t}^2/\tau^2\right)^{1/2}}, \qquad (C.2)$$

where T_0 is the present time background temperature, $T_0 = 2.73$K, $\tilde{t}$ is the cosmic time measured with respect to us now, and τ is the Hubble time in the absence of gravity, $\tau = 12.486$Gyr

[4,5]. Since we are looking for temperatures T at very early times, we can use the approximation $\tilde{t} \approx \tau$, thus

$$1 - \tilde{t}^2/\tau^2 = \left(1 + \tilde{t}/\tau\right)\left(1 - \tilde{t}/\tau\right) \approx 2\left(1 - \tilde{t}/\tau\right)$$

$$= (2/\tau)\left(\tau - \tilde{t}\right) = 2t/\tau, \tag{C.3}$$

where t is the cosmic time with respect to the Big Bang. Using this result in Eq. (C.2) we obtain

$$T = T_0 \left(\tau/2\right)^{1/2} t^{-1/2}, \tag{C.4}$$

with the same dependence on time as in Eq. (C.1).

C.3 Comparison

As is seen, both equations (C.1) and (C.4) show that the temperature T depends on $t^{-1/2}$. The coefficients appearing before the $t^{-1/2}$, however, are not identical. A simple calculation shows

$$\left(\frac{45\hbar^3}{32\pi^3 k^4 G}\right)^{1/4} = 1.52 \times 10^{10} \mathrm{Ks}^{1/2} \tag{C.5}$$

for the coefficient appearing in Eq. (C.1), and

$$T_0 \left(\tau/2\right)^{1/2} = 1.16 \times 10^9 \mathrm{Ks}^{1/2}, \tag{C.6}$$

for that appearing in Eq. (C.4). In the above equations we have used

$$\hbar = 1.05 \times 10^{-34} \mathrm{Js},$$
$$k = 1.38 \times 10^{-23} \mathrm{J/K},$$
$$G = 6.67 \times 10^{-11} \mathrm{m}^3/\mathrm{s}^2\mathrm{Kg}, \tag{C.7}$$
$$T_0 = 2.73 \mathrm{K}, \quad \tau = 12.486 \mathrm{Gyr}.$$

Accordingly we can write for the temperatures in both cases

$$T \approx 1.5 \times 10^{10} \text{Ks}^{1/2} t^{-1/2}, \qquad (C.8)$$

and

$$T \approx 1.2 \times 10^{9} \text{Ks}^{1/2} t^{-1/2}. \qquad (C.9)$$

The ratio between them is approximately 13.

It thus appears that the dominant part of the plasma energy of the early Universe has gone to the creation of matter appearing now in the Universe, and only a small fraction of it was left for the background cosmic radiation.

C.4 References

1. S. Behar and M. Carmeli, to be published.
2. H. Ohanian and R. Ruffini, *Gravitation and Spacetime* (W.W. Norton, New York, 1994) (Eq. 15, Chapter 10).
3. L.D. Landau and E.M. Lifshitz, *The Classical Theory of Fields* (Pergamon Press, Oxford, 1979).
4. M. Carmeli, Basic approach to the problem of cosmological constant and dark energy, in: *Proceedings of the Conference COSMO-01*, held in Rovaniemi, Finland, 29 August – 4 September, 2001. (astro-ph/0111071)
5. M. Carmeli and T. Kuzmenko, Value of the cosmological constant: theory vs. experiment, in: *Proceedings of the 20th Texas Symposium on Relativistic Astrophysics*, held 10 – 15 December, 2000, Austin, Texas, J.C. Wheeler and H. Martel, Editors (American Institute of Physics, 2001). (astro-ph/0102033)

Index

(Page numbers in italics refer to publications cited in the references.)

Aberration,
    angle, 65
    of light, 65
Absolute cosmic time, 3
Absolute rest, 38
Absolute space, 1, 39
Absolute time, 1, 43
Accelerating universe, 128-136, 149, 154-160
Acceleration, 4, 25, 100
    and distances, 102, 103
    minimal, 25, 26
Acceleration four-vector, 4, 75-78, 100-102
Acceleration, velocity and cosmic distances, 99-108
Action-at-a-distance, 44
Addition of cosmic times, 23, 24
Addition of velocities, 4, 41, 60, 61-63
Age of the Universe, 14, 23, 24
Akerlof, C.W., *143, 181*
Angle, aberration, 65
Angular momentum, 91-93,

96, 167
   intrinsic, 96
Angular-momentum representation, 92-94
Angular velocity, 94, 96
Another derivation of the cosmological transformation, 17, 18
Antigravity, 25
Antisymmetric tensor, 72
Apparent incompatibility of the special relativity postulates, 43, 44
Approximate Lorentz transformation, 51
Astronomical object, 9
Asymmetries, 38
Autobiographical Notes (Einstein), 8, *27, 66, 79*

Background, historical, 1, 2
Backward time, 12
Bargmann, V., 35, *36*
Basic concepts, 38
Basic principles of special relativity, 38
Beam, light, 42
Behar, S., *143*, 157, 169, 170, *180, 188*
Behar-Carmeli theory, 170, 171
Bertotti, B., *181*
Bianchi identities, 149, 151
   restricted, 151, 161

Big Bang, 109, 110, 186, 187
Big Crunch, 169
Body, rigid, 42
Bohm, D., *66, 85, 97*
Bond, J.R., *144, 183*
Bondi, H., 44, *66*, 153, *180*
Boomerang experiment, 116, 172
Boost, 51
Born, M., *66, 85, 98*
Branch, D., *181*
Brane world theory, five-dimensional, 145-184
Buchler, J.R., *182*

Carmeli, M., *27, 36, 66, 79, 98, 107, 112, 113, 143*, 157, 169, 170, *180-182, 188*
Carmeli's cosmological trnsformation, 131
Cartan, E., *79*
Christoffel symbols, 155, 176, 177
Circular motion, 167
Circular velocity, 167
Classical field, 39
Classical limit of cosmological transformation, 20
Classical mechanics, 20, 62, 68, 71, 89
   four dimensions in, 68, 69
Classical transformation, 7

Index

inadequacy of, 13, 14
Classical physics, 8, 23
Classification of universes, 125-127
Cline, D., *143*
Clock, 9, 10, 42, 61, 62, 73-75, 78
Clocks and rods, measuring, 42
Cluster, Coma, 22
Comoving coordinates, 120, 151
Comparison with Einstein's special relativity, 111, 112
Composition of cosmic times, 23
Concepts, basic, 38
Cone,
    galaxy, 18-20, 82, 106
    light, 4, 18, 20, 37, 38, 71, 78, 81-85, 106
Consequences of the cosmological transformation, 20-26
Consequences of the Lorentz transformation, 4, 37, 59-65
Conservation law, 87
    covariant, 161
Conservation of energy, 87
Conservation of mass, 87
Constancy of the expansion of the Universe, 4, 9, 10, 29
    principle of, 10, 13
Constancy of the light velocity, 1, 4, 8, 29, 37, 39, 42, 54
    principle of, 4, 10, 13
Constant,
    cosmological, 136, 139, 145, 150, 168-172
    Hubble's, 2, 13, 30, 116, 117, 136, 147, 154
    Planck's, 96
    universal, 147
Contraction of length, 4, 20, 21, 60, 61
Contraction of velocity, 22
Contravariant quantities, 71
Coordinate, time, 41, 43, 55
Coordinate system, 4, 8, 37, 39, 41
    and events, 82-84
    cosmic, 146-148
    inertial, viii, 3, 4, 8, 11, 37, 39-42, 45, 49, 146
Coordinates, 4, 11, 13, 15, 32, 39-43
Cosmic background temperature, 186
Cosmic coordinate systems, 146-148
Cosmic distance, 104, 106
    intrinsic, 104

Cosmic distance in CSR versus energy in ESR, 104
Cosmic distances, acceleration and velocity, 99-108
Cosmic frame, 7, 10-12, 33
Cosmic radiation temperature, 158
Cosmic temperature decline, 185-188
Cosmic time, 2-4, 7, 9-14, 21, 23, 25, 29, 112
    absolute, 3
    relative, viii, 3, 7, 11, 12, 21, 22, 31
Cosmic times, law of addition of, 23, 24
Cosmological constant, 136, 139, 145, 150, 168-172
    Behar-Carmeli predicted value, 170, 171
    supernovae experiments value, 170
Cosmological four-dimensional transformation, 3
Cosmological general relativity, 115-144
Cosmological group, viii, 4, 29, 31, 33
Cosmological redshift, 26, 145, 146, 172-175
Cosmological redshift formula, 160
Cosmological relativity, 3, 7-28, 61, 63
    postulates of, 7, 10
    principle of, 10, 13
Cosmological solutions, 2
Cosmological special relativity, 7-28, 60, 61, 99, 100, 103, 104, 106, 107
    fundamentals of, 99
Cosmological term, 2, 168, 169
Cosmological theory, vii
    five-dimensional, 145
Cosmological transformation, 3, 11, 16-20, 30, 51, 100, 117, 186
    another derivation of, 17, 18
    classical limit, 20
    consequences of, 20-26
    derivation of, 16-18
    interpretation of, 18
    Lorentz-like, 148
Cosmology, vii, viii, 3, 9, 11, 12, 17, 29, 34, 112, 116
    energy density in, 149, 153
    five-dimensional presentation, 154
    present-day, 3, 7, 9
    spacevelocity in, 11
    special relativity of, 3, 7-28
Cosmology and special relativity, 2-4

Index

Cosmology in spacevelocity, 117-120
Cosmos, 3
Covariant conservation law, 161
Covariant field equations, 35
Covariant form, 70
Covariant quantities, 71
Critical density, 169
Curvature 14, 26
Curvature parameter, 168
Curved spacetime, 2

Damour, T., *182*
Dark energy, 175
Dark matter, 36, 167, 172
  remark on, 22
Days, lengths of, 110, 111
De Bernardis, P., *144, 183*
Decline, cosmic temperature, 185-188
Delta, Kronecker, 72
Density of matter, 12, 24
Density parameter, 169
Derivation of the cosmological transformation, 15-18
Derivation of the Lorentz transformation, 44-52
Deruelle, N., *182*
Dilation of time, 4, 22, 59, 61, 74
Dirac equation, 35

Disk galaxies, 167
Distance, 3, 10, 33
  longest, 34
Distance scales,
  'long', 129, 160
  'short', 129, 160
Distance-velocity four-vector, 104-106
Distances and acceleration, 102, 103
Distribution of galaxies, 18
Dual space, 2, 3, 18
Dynamical variables, 2, 33

Earth, 28, 41, 45, 46
Effective mass density in cosmology, 153, 154
Effective potential, 166, 167
Einstein, A., 1, 2, *5*, 8-10, *28*, 40, 45, 58, 61, *71*, *86*, *105*, *113*, *143*, 152-154, 162, 168, *180*, *181*
Einstein-Infeld-Hoffmann method, 162
Einstein's Autobiographical Notes, 8, *27, 66, 79*
Einstein's famous formula, 90
Einstein's gravitational field equations, 2
Einstein's paradox, 42, 43
Einstein's relativity, vii
Einstein's special relativity, viii, 3, 4, 7, 99, 103, 106,

112
    comparison with, 111, 112
Einstein tensor, 119
Eisenhart, L.P., *180*
Electrodynamics, 38
    of moving charges, 1
Electromagnetic field, 38, 39, 42
Elementary particles, vii
Energy, 4, 38, 106, 107
    conservation of, 87
    kinetic, 88, 93
    rest, 88, 106
Energy and mass, relationship between, 99
Energy-angular-momentum formula, 94
Energy density,
    in cosmology, 149, 153
    vacuum, 26
Energy in ESR versus cosmic distance in CSR, 104
Energy, mass and momentum, 87-98
Energy-momentum, 4
Energy-momentum formula, 92, 94
Energy-momentum four-vector, 4, 38, 95
    in ESR, 104
Energy-momentum tensor, 119, 168
Equation,

Dirac, 35
Freidmann, 169
geodisic, 163-166
Klein-Gordon, 35
Proca, 35
Weyl, 35
Equations of Maxwell, 35, 43, 79
Equations of motion, 163-166
Equivalence of energy and inertial mass, 90
Erez, G., viii
Essence of special relativity, 8, 43
Ether, 21
Event, 23, 40, 42, 44, 57, 69, 78, 84, 87
Events and coordinate systems, 82-84
Events, simultaneous, 40
Evolution,
    three-phase, 170
Expanding universe, 2, 3, 147, 156-158
    equations of motion, 149
Expansion, Universe, 4, 7, 10, 13, 14, 29, 31, 32, 147, 169, 175, 178, 179
    accelerating, 158, 175, 176, 180
    constant, 158, 180
    decelerating, 158, 176
    equation integration, 178-

180
Experiment of Michelson and Morley, 38, 43
Extension of the Lorentz group to cosmology, 29-36

Field, 44
    classical, 39
    electromagnetic, 38, 39, 42
    gravitational, vii, 4, 37
Field equations, 170
    covariant, 35
    Einstein, 119, 150-152, 156
    Einstein's gravitational, 2
    gravitational, 120-122, 151, 152
    linear, 35
    solution of, 123, 124
Figure 2.1, 19
Figure 4.1, 48
Figure 5.1, 76
Figure 6.1, 83
Figure A.1, 130
Figure A.2, 132
Figure A.3, 133
Figure A.4, 134
Figure A.5, 135
Figure A.6, 140
Figure A.7, 141
Filippenko, A.V., *144*, *183*
Finite-mass particle, 63
First days of the Universe, 109-114
Five-dimensional brane world theory, 145-184
Five-dimensional manifold of space, time and velocity, 149, 150
Flat-space metric, 45, 71
Fock, V., 153, *181*
Force, Newtonian, 163
Formula,
    cosmological redshift, 160
    energy-angular-momentum, 94
    energy-momentum, 92, 94
    redshift, 172-175
    temperature, 186, 187
    thermodynamical, 186
    Tully-Fisher, 145, 149, 160-168
Foundations of special relativity, 7, 67
Four-dimensional continuum, 68
Four-dimensional cosmological transformation, 4, 15-18, 21, 33
Four-dimensional rotation, 4, 31
Four-dimensional rotation group, 4, 29, 31, 33
Four-dimensional spacetime, 67, 78
Four-dimensional structure of

spacetime, 63, 78
Four dimensions in classical mechanics, 68, 69
Four-quantities, 40
Four-tensor, 70
Four-vector, 4, 70, 71
    acceleration, 4, 75-78, 100-102
    distance-velocity, 104-106
    energy-momentum, 4, 38, 95, 104
    interval, 82, 84
    momentum, 87
    position, 70, 71, 91
    velocity, 4, 40, 75-78, 95, 100-102
Frame,
    cosmic, 7, 10, 11, 31
    inertial, 30
    moving, 10
    reference, 12
    stationary, 8
Frames,
    inertial, 1, 11
Freedman, W.L., *144, 181*
French, A.P., *67, 79, 85, 97*
Friedmann, A., 2, *5*, 170
Friedmann equations, 168, 169
FRW model, 175
Fundamentals of special relativity, 8, 37-67, 89, 95
Future, 85

Future and past, 84, 85

Galaxy cone, 18-21, 82, 106
Galaxies, 2, 14, 18, 22
    disk, 167
    distribution of, 18
    locations of, 18, 20
    receding, 2
    spiral, 22
Galaxy,
    spherically-symmetric, 145
Galilean,
    group, 4, 37, 41
    invariance, 41
    transformation, 1, 4, 11-13, 37, 40, 41, 43, 52, 59, 60, 67, 146, 148
Garnavich, P.M., *144, 182*
Gas molecules, 22
General relativity, cosmological, 115-144
General relativity theory, 2, 118, 119, 154, 161
Generalized Minkowskian metric, 32
Generalized transformation, 34, 35
Geodesic equation, 161-163
Goldstein, H., *182*
Gottesman, S.T., *182*
Gradshteyn, I.S., *143*
Gravitation, 1, 3
Gravitational field, vii, 37

Index

Gravitational field equations, 120-122, 151, 152
   Einstein's, 2
Gravity, 11
Grommer, J., *181*
Group,
   cosmological, viii, 4, 29, 31, 33, 36, 149
   four-dimensional rotation, 4, 29, 31, 33, 36
   Galilean, 4, 37, 41
   homogeneous Lorentz, 3, 4, 29, 52
   inhomogeneous Lorentz, 35, 52
   Lorentz, viii, 4, 29, 31, 33, 37, 52-54, 149
   orthochronous Lorentz, 53
   O(2), 149
   O(4), 31
   O(1,3), 31, 69, 149
   O(2,3), 149
   O(3,1), 149
   O(3,3), 31
   O(3,4), 31, 35
   Poincaré, 52
   proper Lorentz, 53, 54
   two-dimensional Euclidean, 149
Group of cosmological transformations, viii
Group of transformations, 149
Group representations, 35, 36, 52
Guth, A.H., *27*

Halo dark matter, 145, 160-168
   around galaxies, 149
High-energy physics, vii
High velocities, 22
High-Z Supernova Team, 116, 137, 145, 170, 172
Higher-dimensional space, 3
Highest velocity, 34
Highway patrol, 10
Historical background, 1, 2
Hoffmann, B., 162, *181*
Homogeneous and isotropic universe, 14
Homogeneous Lorentz group, 3, 4, 29, 52
Hubble, E., viii, 2, *5*, 115, 169
Hubble's constant, 2, 13, 30, 116, 117, 147, 154
Hubble expansion, 117, 156
Hubble's law, 2, 3, 10, 11, 13, 15, 29, 116-118, 153
   consequences of, 3
Hubble parameter, 169, 172
Hubble space telescope, 142, 144, 181
Hubble's time, 9, 12, 14, 23, 30, 100, 109-111, 116-119, 149

Hubble transformation, 146-148
Hubble's variables, 12
Hunter, J.H., *182*
Hydrogen atoms, 26, 120

Improper Lorentz transformation, 52
Inadequacy of classical transformation, 13
Indices, lower and upper, 71
Inertia, law of, 39
Inertial coordinate system, viii, 3, 4, 8, 11, 37, 39-45, 48, 68, 73, 74, 81, 146
Inertial frame, 30
Inertial frames, 1, 11, 30
Inertial mass, 90
Infeld, L., 162, *181*
Infinite-dimensional representations, 36
Inflation of space, 3
Inflation of the Universe, 24, 25
Inflationary universe, vii, 3, 25, 27
Inhomogeneous Lorentz group, 35, 52
Inhomogeneous Lorentz transformation, 52
Integration of the Universe expansion equation, 178-180
Interpretation of the cosmological transformation, 17
Interval four-vector, 82, 84
Intrinsic angular momentum, 96
Invariance, Galilean, 41
  Lorentz, 69, 88
Invariant, 68
  law of physics, 68
  Lorentz, 71
Inversion, space, 53

Karade, T.M., *143*
Kinetic energy, 88, 93
Klein-Gordon equation, 35
Kronecker delta, 72
Kuzmenko, T., *107, 113, 188*

Landau, L.D., *66, 79, 85, 182, 188*
Large-scale structure of the Universe, vii, 3
Law,
  conservation, 93
  Hubble's, 2, 3, 11, 13, 15, 29, 116-118, 153
  physical, 70
  Tully-Fisher, 166-168
Law of addition of cosmic times,

Index 199

3, 23, 24
Law of addition of velocities, 41, 61-63
Law of composition of cosmic times, 23
Law of inertia, 39
Laws, physical, 1
Length, 20, 21, 60
    proper, 21
Length contraction, 4, 20, 21, 60
Lengths of days, 110, 111
Lifshitz, E. M., *66, 79, 85, 182, 188*
Light, 1, 14, 26, 29, 42-44
    aberration of, 65
    speed of, 14, 30, 38, 40, 42, 45, 63, 111, 147
    propagation of, 147
    wave length of, 26
Light aberration, 69, 70
Light beam, 42
Light cone, 4, 18, 38, 71, 78, 81-88, 106
Light propagation, 1, 2, 7, 9, 12, 13, 44, 45, 67, 148
    versus Universe expansion, 13, 14
Light pulse, 44
Light ray, 54, 55
Light signal, 40, 42, 57, 82
Light speed, 1, 13, 30, 39, 40, 42, 45, 63, 96, 118, 119
Light velocity,
    constancy of, 4, 8, 29, 42, 54, 88
Limiting velocity, 63
Linde, A.D., *27*
Line, world, 72, 73, 75
Line element, 24, 32
    cosmological flat spacevelocity, 148
    flat spacetime, 148
    in five dimensions, 156
Linear,
    field equations, 35
    transformation, 45
Linear field equations, 35
Local (noncosmological) physics, 11
Local time, 42
Locations of galaxies, 18, 20
Longest distance, 34
Lorentz, H.A., *180*
Lorentz contraction, 21, 40, 60
Lorentz contraction factor, 91, 92
Lorentz group, viii, 3, 4, 29, 31, 33, 38, 52-54
    extension to cosmology, 29-36
    homogeneous, 3, 4, 29, 52
    inhomogeneous, 35, 52

orthochronous, 54
proper, 54
Lorentz invariance, 69, 78
Lorentz invariant, 72
Lorentz-like cosmological transformation, 148, 149
Lorentz transformation, 1, 3, 4, 8, 9, 11-13, 17, 37, 42-59, 62, 63, 67-70, 89, 100, 117, 148
    approximate, 51
    consequences of, 4, 37, 59-65
    derivation of, 44-52
    extension of, 3
    homogeneous, 4
    improper, 52, 54
    orthochronous, 53
    proper, 52, 54
Luminoferous ether, 22
Luminosity, 167

Malin, S., *66*
Manifold, 68, 71
Manset, N., *112*
Martel, H., *107, 113, 188*
Mass, 4, 38
    inertial, 90, 100, 103
    rest, 90, 95, 100, 103, 106
Mass and energy, relationship between, 99
Mass density, 119, 169
    critical, 120

    effective, 153, 154
Mass, energy and momentum, 87-98
Matrix, orthogonal, 46
    unit, 71
Matter,
    dark, 23, 36
    density of, 12, 24, 26
Maximum signal speed, 40
Maximum time, 34
Maximum velocity, 12
Maxwell's equations, 35, 43
Maxwell's theory, 39
Measurement, 42
Measuring rods and clocks, 42
Mechanics, 20, 38, 39, 41, 62, 71, 75, 77, 89
Metric,
    Minkowskian, 32, 34, 177
    flat-space, 45, 71
Michelson and Morley experiment, 38, 43
Milgrom, M., 26, *27*
Miller, A.I., *66*
Minimal acceleration, 25, 26
Minkowski, H., 1, 4, *5*, 37, 67, *78, 85, 180*
Minkowskian metric, 34
    generalized, 32
Minkowskian spacetime, 14, 46, 69-72, 81
Minkowskian spacetime dia-

Index

gram, 75
Modern physics, 44
Molecules, gas, 22
Moment of inertia, 93, 96
Momentum, 5, 38, 90, 91
    angular, 91-93, 167
Momentum, mass and energy, 87-98
Momentum four-vector, 87
Moving frame, 10

Ne'eman, Y., viii
Negative timelike vector, 71
Newton, 136
Newton's laws of motion, 41, 167
Newtonian force, 163
Newtonian gravity, 119
Newtonian kinetic energy, 88
Newtonian mechanics, 39, 41, 89, 165, 166
Nieves, J.F., *144, 183*
Nonrelativistic limit, 59, 60
Nonstatic cosmological solutions, 2
Null condition, 116, 118, 120
Null propagation of light, 14
Null vector, 14, 71

Observable measured quantities, 3
Observer, viii, 3, 10, 22, 26, 30, 42, 45, 48, 74
Ohanian, H., *182, 188*
Orthochronous Lorentz group, 53
Orthochronous Lorentz transformation, 53
Orthogonal matrix, 46
Orthogonal rotation, 40
Orthogonal transformation, 46
$O(4)$ group, 31
$O(1,3)$ group, 31, 69
$O(3,3)$ group, 31
$O(3,4)$ group, 31, 35
$O(3) \times T(1)$, 68

Paradox,
    Einstein's, 42, 43
    twins, 74
Particle, 11, 46, 63, 72, 81, 90, 92-96
    test, 161
Particles,
    decay of, 74
    elementary, vii
Past, 85
Past and future, 84, 85
Peebles, P.J.E., *143, 181*
Periodic process, 42
Perlmutter, S., *144, 182, 183*
Perpetuum mobile, 9
Photon, 88, 93
Physical consequences, 3
Physical interpretation of spa-

tial coordinates and time, 42
Physical law, 70
Physical laws, 1
Physics, vii, 7, 43, 44, 68, 69
    classical, 23
    high-energy, vii
    local (noncosmological), 11
    prerelativistic, 3, 8
Pierce, M., *181*
Piran, T., *182*
Planck's constant, 96
Plebanski, J., *181*
Poincaré group, 52
Position four-vector, 70, 71, 91
Positive timelike vector, 71
Postulates of cosmological special relativity, 7, 10
Postulates of special relativity, 4, 37-40, 43, 67
Potential, effective, 166, 167
Preliminaries, 29, 88
Prerelativistic physics, 3, 8
Prerelativistic times, 9
Prerelativity, 11, 12, 42
Present-day cosmology, 3, 7, 9
Pressure, 116
Principle of constancy of the speed of light, 4, 10, 13
Principle of cosmological relativity, 10, 13
Principle of equivalence, 118
Principle of general covariance, 118
Principle of special relativity, 4, 8-10, 13, 38, 43, 44, 60, 67
Principle of the constancy of the propagation of light, 10, 13, 37
Proca equation, 35
Process, periodic, 42
Product, scalar, 71
Propagation, light, 1, 2, 7, 9, 12, 13, 44, 45, 73, 147
Proper length, 21
Proper Lorentz group, 53
Proper Lorentz transformation, 52, 54
Proper orthochronous Lorentz group, 53
Proper time, 4, 40, 72-75
Pulse, light, 44

Radar devices, 10
Radial velocity, 3
Radius of gyration, 96
Ray, light, 54, 55
Receding galaxies, 2
    velocities of, 117
Receding velocity, 2, 14
Redshift, 17, 26, 27, 174

Index

cosmological, 26, 27, 145, 146, 172-175
Redshift analysis, 150
Redshift formula, 172-175
Redshift parameter, 129
Reference frame, 12
Relative cosmic time, viii, 3, 7, 11, 12, 21-23, 31
Relative velocity, viii, 3, 12
Relativity,
    cosmological general, 115-144
    cosmological special, 3, 7-28, 60, 61, 63
    general, 2, 162
    principle of, 4, 8, 10, 13, 37, 38, 43, 60
    special, vii, viii, 2-4, 7, 10, 12, 17, 21, 22, 24, 27, 38, 44, 60, 61, 68, 69, 93, 94, 97
Relativity theory, vii
Representations of group, 35, 52
Rest energy, 88
Rest mass, 90, 95
Rest moment of inertia, 93
Ricci scalar, 119, 151, 155, 178
Ricci tensor, 119, 151, 155, 178
    components of, 177, 178
    tracefree, 119

Riess, A.G., *144, 181-183*
Rigid body, 42
Rod, 9, 23, 42, 43, 61
Rods and clocks, measuring, 42
Rotation,
    four-dimensional, 4, 31
    orthogonal, 40
    three-dimensional, 34, 51
Rotation group,
    four-dimensional, 4, 29-31, 33
Ruler, 10
Ruffini, R., *182, 188*
Russull, Bernard, 8
Ryshik, I.M., *143*

Saha, A., *181*
Sandage, A., *181*
Scalar, 70
    Ricci, 119, 151, 155, 178
Scalar product, 71
Scale factor, 168
Schild, A., *181*
Schmidt, B.P., *144, 181, 182*
Schwarzschild radius, 174
Signal, light, 40, 42, 57
Signal speed, maximum, 40
Signature, 32
Simpson, C., *112*
Simultaneity, 4, 37, 40
Skew-symmetric tensor, 72
SNe Ia, 139

Solution of the field equations, 123, 124
Solutions, cosmological, 2
Space, vii, viii, 1-4, 8, 9, 14, 68
   absolute, 1, 39, 154
   and time, 2
   and velocity, 2
   cosmological, 154
   dual, 2, 3, 18
   Euclidean, 154
   inflation of, 3
Space inversion, 53, 54
Space telescope, Hubble, 142
Space, time and velocity, five-dimensional manifold of, 149, 150
   unification of, 161
Spacelike, 71
Spacetime, 1, 3, 37
   four-dimensional, 67, 78, 81
   Minkowskian, 14, 46, 69-72, 81
   structure of, 3, 63, 67-80
Spacetime symmetry, 38
Spacevelocity, 7, 26
   in cosmology, 11, 117-120
   structure of, 3
Spacevelocity analysis, four-dimensional, 100
Spatial coordinates, 3, 8, 39, 41-45, 51, 67, 69
   physical interpretation of, 42
Special relativity, vii, viii, 1-4, 7, 10, 12, 17, 21, 22, 24, 27, 38, 40, 43, 44, 62, 63, 74, 75, 93, 94, 97, 107, 116, 117
   and cosmology, 2-4
   basic principles of, 38
   comparison with, 111, 112
   cosmological, 7-28, 60, 61, 99, 100, 103, 104, 106, 107
   Einstein's, 99, 103, 106, 112
   essence of, 8, 43
   foundations of, 7, 67
   fundamentals of, 8, 37-68, 87, 95
   postulates of, 4, 37-40, 43, 67
   principle of, 4, 8-10, 13, 37, 38, 43, 44, 67
Special relativity of cosmology, 3, 7-28
Special relativity as a valuable guide, 68
Speed of light, 1, 4, 13, 14, 30, 39, 40, 42, 44, 63, 96, 111, 118, 119, 147
Spinor, 70
Spiral galaxies, 22
Srednicki, M.A., *143, 181*

Index

Static universe, 2
Stationary frame, 8
Strokes, 61
Structure of,
    spacetime, 3, 63, 67-80
    spacevelocity, 3
Supernova Cosmology Project, 116, 137, 145, 170
Supernova SN 1997ff, 175, 176
Supernovae, 150
    type Ia, 170, 175
Symmetric tensor, 72
Symmetry, spacetime, 38
between R and $-\kappa T$, 151
System,
    coordinate, 4, 9, 37, 39-41
    inertial coordinate, viii, 3, 4, 8, 11, 37, 39-45, 49, 68

Table A.1, 131
Table A.2, 142
Temperature, 12
    cosmic, 185-188
    cosmic background, 186
    Universe, 185
Tensor, 70, 72
    Einstein, 119
    energy-momentum, 119, 120, 168
    Ricci, 119, 151, 155, 177, 178
    Ricci tracefree, 119
Term,
    cosmological, 168, 169
    Tully-Fisher, 167
Test particle,
    spherically symmetric, 161
Thermodynamical formula, 186
Thermodynamics, 2, 9
Three-dimensional rotation, 51
Three-phase evolution, 170
Three-vector, 71
Ticks, 61
Time, vii, viii, 1-4, 8, 9, 29, 33, 40, 42, 44, 45, 60, 67-69
    absolute, 1, 43
    absolute cosmic, 3
    and space, 2
    backward, 12
    cosmic, 2-4, 7, 9-14, 17-19, 21, 23-26, 29, 112
    dilation of, 4, 22, 40, 59, 61, 74
    Hubble's, 9, 12, 14, 23, 100, 109-111, 116-119, 149
    local, 42
    maximum, 34
    physical interpretation, 42
    prerelativistic, 10
    proper, 4, 72-75
    relative cosmic, viii, 3, 8, 11, 12, 21-23, 31

upper limit of, 12
Time coordinate, 39, 40, 51
Time lengths of the first days, 109
Time reversal, 53
Timelike, 71
Times, addition of cosmic, 23, 24
Trace, 72
Transformation,
   approximate Lorentz, 51
   classical, 7, 13
   cosmological, 3, 11, 15-18, 20, 30, 51, 100, 117, 148, 149, 186
   four-dimensional cosmological, 3
   Galilean, 1, 4, 11-13, 37, 40, 41, 43, 52, 59, 60, 67, 146, 148
   generalized Minkowskian, 32
   Hubble, 146-148
   improper Lorentz, 52, 54
   linear, 45
   Lorentz, 1, 3, 4, 8, 9, 11-13, 17, 37, 41-59, 62, 63, 67-70, 100, 117, 148
   orthochronous Lorentz, 53
   orthogonal, 46
   proper Lorentz, 52-54
   spacevelocity, 105

Transformations, group of cosmological, viii
Translation, 40, 52
Tully-Fisher formula, 145, 149, 160-168
Tully-Fisher law, 166-168
Tully-Fisher term, 167
Turner, M.S., *183*
Twins paradox, 74

Unification of space and time, 3
Unification of space and velocity, 3
Unification of space, time and velocity, 161
Uniform motion, 40
Unit matrix, 71
Universe, vii, viii, 2-4, 7, 10, 11, 18, 23, 25, 115, 116, 118, 119, 188
   accelerating, 115, 120, 128-136, 149, 150, 154-160, 170
   age of, 14, 23, 24, 111
   constant, 150
   decelerating, 150
   early, 109, 110, 188
   Euclidean, 138, 172
   expanding, 2, 3, 120, 147, 149, 156-158, 169
   first days of, 109-114
   homogeneous and isotropic,

Index

14, 168
in five dimensions, 145
infinite, 150
inflation of, 24, 25
inflationary, vii, 3, 25, 27
large-scale structure of, vii, 3
open, 146, 150, 160, 175
static, 2
temperature of, 12
three-phase evolution, 115
with gravitation, 150-154
Universe expansion, vii, 7, 10, 13, 14, 29, 116, 120
versus light propagation, 13, 14
Universes, classification of, 125-127
Upper limit of time, 12

Vacuum, 2
Vacuum energy density, 26
Van Dam, H., 44, *66*
Variables, dynamical, 2, 33
Vector, null, 14
Velocities, high, 22
addition of, 4, 41, 59, 61-63
Velocities of galaxies, receding, 117
Velocity, vii, viii, 2, 3, 9, 17, 24, 33, 91, 118
and space, 3

angular, 94, 96
as independent coordinate, 149, 152, 153
circular, 167
highest, 34
limiting, 63
maximum, 12
receding, 2, 14
relative, viii, 3, 12
three-dimensional, 103, 105
Velocity, acceleration and cosmic distances, 99-108
Velocity contraction, 22, 61
Velocity four-vector, 4, 40, 75-78, 95, 100
in spacevelocity, 101
Velocity measuring instrument, 22
Velocity of light, 63
Von Hippel, T., *112*

Wave length, 26
Weyl, H., *180*
Weyl equation, 35
Wheeler, J.C., *107, 113, 188*
Whitemore, B.C., *182*
Wigner, E.P., 34, *35*, 44, *66*
World line, 72, 73, 75